Style in Technical Math
Solution Manual

Afshin Azari-Vala

An Overview of the Manual

This publication provides detailed solutions for the problems in the exercise sets in the textbook *Style in Technical Math* by the author. The coverage includes both even-numbered and odd-numbered problems in the exercise sets.

As our aim is to promote formulations and algorithms that promote fluency, for the most part, we have provided detailed solutions as opposed to partial solutions or just answers. Alternative approaches that meet the criteria for fostering fluency in solving mathematical problems are included in the solutions.

When an exercise extends a concept that is introduced in the body of the textbook or introduces a new one, further detail about the topic is included in the form of side notes. This should help the reader connect the ideas that are presented in the body of the text to their extension in the exercises.

Ideally, this manual should be consulted as a check on solutions that the reader provides. While study skills differ, it is often the case that active engagement through solving problems *using pen and paper* is much more effective than passive participation through looking up the solutions to problems that are posed for practice.

Solutions to Exercise Sets

Exercise Set 1

1. To answer these questions, take the following steps:

    i. Simplify the given expression.

    ii. Find the smallest set that the simplified form belongs to. Does the simplified form belong to $\mathbb{N}$? $\mathbb{W}$? $\mathbb{I}$? $\mathbb{Q}$ (or is it indeterminate or undefined)? $\mathbb{R}$ (or is it indeterminate or undefined)? Stop as soon as you find a set that the simplified form belongs to. This is the smallest set that the given number belongs to.

    iii. Once you have found this set, use Figure 1.5.1 in the textbook to determine what other sets the given number belongs to.

    As an example, the expression $\frac{3}{4}$ is not a natural number. Nor is it a whole number or an integer. It is a rational number, however, and this makes $\mathbb{Q}$ the smallest set that the number $\frac{3}{4}$ belongs to. As shown in Figure 1.5.1, $\mathbb{Q}$ is inside $\mathbb{R}$. This means that the number $\frac{3}{4}$ also belongs to $\mathbb{R}$.

    As a second example, the expression $\sqrt{16}$ simplifies to 4. The smallest set that the simplified form 4 belongs to is $\mathbb{N}$. This means that $\mathbb{N}$ is the smallest set that $\sqrt{16}$ belongs to. According to Figure 1.5.1, this implies that $\sqrt{16}$ also belongs to $\mathbb{W}, \mathbb{I}, \mathbb{Q}$ and $\mathbb{R}$.

    a. $\mathbb{N}, \mathbb{W}, \mathbb{I}, \mathbb{Q}, \mathbb{R}$

    b. $\mathbb{I}, \mathbb{Q}, \mathbb{R}$

    c. This expression is indeterminate, i.e., it can equal to any number.

    d. $\mathbb{Q}, \mathbb{R}$

    e. $\mathbb{R}$

    f. $\mathbb{W}, \mathbb{I}, \mathbb{Q}, \mathbb{R}$

    g. This expression is undefined, i.e., it can not equal to any numbers.

    h. $\mathbb{N}, \mathbb{W}, \mathbb{I}, \mathbb{Q}, \mathbb{R}$

    i. $\mathbb{I}, \mathbb{Q}, \mathbb{R}$

    j. $\mathbb{W}, \mathbb{I}, \mathbb{Q}, \mathbb{R}$
       Note that $\frac{0}{9}$ simplifies to 0.

1

k. $\mathbb{Q}$, $\mathbb{R}$

l. $\mathbb{R}$

m. $\mathbb{N}$, $\mathbb{W}$, $\mathbb{I}$, $\mathbb{Q}$, $\mathbb{R}$
   Note that $\frac{12}{3}$ simplifies to 4.

n. $\mathbb{N}$, $\mathbb{W}$, $\mathbb{I}$, $\mathbb{Q}$, $\mathbb{R}$
   Note that $\sqrt{9}$ simplifies to 3.

o. $\mathbb{Q}$, $\mathbb{R}$

p. $\mathbb{N}$, $\mathbb{W}$, $\mathbb{I}$, $\mathbb{Q}$, $\mathbb{R}$

q. $\mathbb{I}$, $\mathbb{Q}$, $\mathbb{R}$

r. This expression is undefined, i.e., it can not equal to any numbers.

s. $\mathbb{Q}$, $\mathbb{R}$

t. $\mathbb{R}$

u. $\mathbb{N}$, $\mathbb{W}$, $\mathbb{I}$, $\mathbb{Q}$, $\mathbb{R}$
   Note that $\frac{15}{5}$ simplifies to 3.

v. $\mathbb{Q}$, $\mathbb{R}$

w. $\mathbb{I}$, $\mathbb{Q}$, $\mathbb{R}$
   Note that $-\sqrt{16}$ simplifies to -4.

x. $\mathbb{I}$, $\mathbb{Q}$, $\mathbb{R}$

y. $\mathbb{W}$, $\mathbb{I}$, $\mathbb{Q}$, $\mathbb{R}$
   Note that $\sqrt{0}$ simplifies to 0.

2. a. The set of integers
   b. The set of rational numbers
   c. The set of real numbers
   d. The set of natural numbers
   e. The set of irrational numbers
   f. The set of whole numbers

3. a. Using mathematical notation we can describe the set of rational numbers as

$$\mathbb{Q} = \left\{ \frac{p}{q} \mid p \wedge q \in \mathbb{I} \wedge q \neq 0 \right\}$$

   We can also use English to describe the set as follows: The set of rational numbers is the set of all numbers that have the form of $\frac{p}{q}$ where p and q are integers and q is not 0.

   b. Expressions that involve division by 0 are either of the form $\frac{0}{0}$ which is indeterminate (meaning it can equal to any number) or they have the form $\frac{n}{0}$ with $n \neq 0$ which are undefined (they do not equal to any numbers). As such, neither expression qualifies as a specific type of number.[1]

   c. The word *rational* derives from the word *ratio* which in turn relates to the comparison of the values of two similar quantities by division.

---

[1] See Appendix A in the textbook for more on indeterminate and undefined expressions.

4. a. A real number that cannot be written as a rational number.
   b. There are no numbers that are both rational and irrational. Note that *irrational* means *not rational* and as such, an irrational number can not be rational.

5. To answer these questions, use Figure 1.5.1 in the textbook.
   a. True. 0 is a whole number but not a natural number.
   b. True. Every integer can be placed over 1 to assume the form of a rational number.
   c. True. The set of real numbers is an extension of the set of rational numbers. Therefore, every rational number is a real number.
   d. False. While some rational numbers reduce to an integer (e.g., $\frac{-12}{3}$ which reduces to -4), others do not (e.g., $\frac{3}{4}$).
   e. False. The set of real numbers is made up of all rational numbers and all irrational numbers grouped together. As such, all irrational numbers are real.
   f. True. Every natural number can be placed over 1 to assume the form of a rational number.
   g. True. The set of integers is an extension of the set of whole numbers. Therefore, every whole number is an integer.

6. a. $\mathbb{N}, \mathbb{W}, \mathbb{I}, \mathbb{Q}, \mathbb{R}$
   b. $\mathbb{I}, \mathbb{Q}, \mathbb{R}$
   c. $\mathbb{N}, \mathbb{W}, \mathbb{I}, \mathbb{Q}, \mathbb{R}$
   d. None[2]

7. It means that the result of multiplying two integers is an integer.

8. a. This tells us that the result of adding two natural numbers is a natural number and therefore, when solving problems that involve natural numbers and addition, we can be sure that there will be notation available for us to write, interpret and work with the result.
   b. This tells us that the result of subtracting two integers is an integer and therefore, when solving problems that involve integers and subtraction, we can be sure that there will be notation available for us to write, interpret and work with the result.

9. a. It means that the given expression can equal to any number.
   b. It means that the given expression is meaningless in the sense that it cannot equal to any numbers.

---

[2] We could, of course, say that the set of rational numbers is closed under division *other than division by* 0. This can be formally stated as follows: The set $\mathbb{Q} - \{0\}$, i.e., the set of rational numbers excluding 0 (and, incidentally, the set $\mathbb{R} - \{0\}$, i.e., the set of real numbers excluding 0) are closed under division.

Exercise Set 2

1. a. The strengths of the formal notation are in being exact and descriptive. Its weakness is that it is a difficult notation to work with.

   b. The strength of base-ten notations is that they are easier to work with. Their weaknesses are that they are not always exact and are not as descriptive.

2. The decimal notation and scientific notation.

3. a. The formal notation should be used in the expression of scientific theories which are assumed to be exact and required to be descriptive.

   b. Base-ten notations are suitable for the expression of the numerical values of quantities in the applications of scientific theories. Such numerical values are inherently approximate and lack significance.

Exercise Set 3

1. The formal notation. The reason is that theoretical discussions are assumed to be exact and required to be clear. These demands are nicely met by the exact nature of the formal notation and its descriptiveness.

2. Symbols are used to refer to the *exact* value of a constant in a science formula. The use of measured numerical values for constants would make science formulas approximate as measured values are rarely exact.

3. The appearance of a numerical value in its formal form in a science formula implies that the value is exact and is based on theoretical considerations.

4. The use of formal notation in the communication of science ideas indicates that the discussion is theoretical. Appearance of base-ten notations signals the shift from theory to applications of theory.

Exercise Set 4.1.1

1. a. thousands triad
   b. ones triad (in practice this label is dropped)
   c. millions triad
   d. billions triad

2. a. ones or units   c. thousands   e. hundred thousands
   b. ten millions   d. hundreds

3. tens   6. ones or units   9. hundred thousands
4. thousands   7. ten thousands
5. hundreds   8. millions

10. a. 40 + 5
       4 tens and 5 units

    b. 9000 + 300 + 70 + 1
       9 thousands, 3 hundreds, 7 tens and 1 unit

c. $4000 + 20 + 7$
4 thousands, 2 tens and 7 units

d. 400
4 hundreds

e. $20\,000 + 3000 + 400 + 20$
2 ten thousands, 3 thousands, 4 hundreds and 2 tens

f. $9\,000\,000 + 200\,000 + 5000 + 60 + 9$
9 millions, 2 hundred thousands, 5 thousands, 6 tens and 9 units

11. The place value of a given digit is 10 times larger than the place value of the digit on its right.
12. The place value of a given digit is 1000 times larger than the place value of the third digit on its right.
13. The place value of a given digit is 100 times smaller than the place value of the second digit on its left.

14. A *hundred* is 10 times larger than a *ten*.
15. A *ten thousand* is 100 times larger than a *hundred*.
16. A *million* is 1000 times larger than a *thousand*.
17. A *ten* is 10 000 times smaller than a *hundred thousand*.
18. A *hundred* is 10 000 times smaller than a *million*.
19. A *ten million* is 100 times larger than a *hundred thousand*.
20. 100 *tens* fit in a *thousand*.
21. 10 000 *ones* fit in a *ten thousand*.
22. 1000 *thousands* fit in a *million*.
23. 10 *ten thousands* fit in a *hundred thousand*.
24. 1 *hundred* fits in a *hundred*.

25. If the digit is in the rightmost triad, use *ones*, *tens* or *hundreds* depending on the location of the digit within the rightmost triad. If the digit is in a triad other than the rightmost triad, note the name of the triad the digit is in and precede the name with blank, *ten* or *hundred* depending on the relative location of the digit within the triad.

Exercise Set 4.1.2

1. a. twenty-five
   twenty-fifth

   b. three
   third

   c. fourteen
   fourteenth

   d. four thousand
   four thousandth

   e. seven thousand one hundred
   seven thousand one hundredth

    f. four thousand two hundred eighty-nine
      four thousand two hundred eighty-ninth

    g. thirty-two thousand four hundred ten
      thirty-two thousand four hundred tenth

    h. two thousand four hundred nineteen
      two thousand four hundred nineteenth

    i. thirty-six
      thirty-sixth

    j. two hundred thirty million one hundred ninety-six
      two hundred thirty million one hundred ninety-sixth

    k. four hundred thirty-eight thousand
      four hundred thirty-eight thousandth

    l. three hundred twenty-four thousand twenty
      three hundred twenty-four thousand twentieth

2. a. 46   c. 350   e. 45 000   g. 73 526   i. 2 500 070
   b. 18   d. 8527   f. 9400   h. 215   j. 55 000 001

Exercise Set 4.1.3

1. a) $36 < 74$ c) $350 > 200 > 178$
   b) $15 > 2$ d) $241 < 320 < 324 < 418$

2. $4, 24, 620, 1242$ 4. $560, 241, 54, 40, 38$
3. $4521, 4762, 5416, 5817$ 5. $3729, 3109, 2410, 420, 197$

Exercise Set 4.1.4

1. a. 37×88 i. $12 \times 11 = 132$
   b. $\frac{32}{4} = 8$ j. $36 + 70 = 106$
   c. $182 + 16 + 4120$ k. $410 + 9100 + 3162 = 12\,672$
   d. 96×79 l. $182 - 46$
   e. $15 + 23$ m. $67 - 52$
   f. $4200 - 3190$ n. $3146 - 2050 = 1096$
   g. $416 - 210$ o. $18 \times 4 = 72$
   h. $\frac{26}{13}$ p. $\frac{210}{15}$

2. a. 569   f. 1412   k. 738   p. 15   u. 32 R1
   b. 6891   g. 69 899   l. 62 000   q. 302   v. 20
   c. 86   h. 3914   m. 760 000 r. 1800   w. 12 R10
   d. 210   i. 21 480   n. 3400   s. 26   x. 360
   e. 1113   j. 998   o. 67   t. 15 R4

3. 8   5. 123 328   7. 12   9. 5155   11. 24
4. 142   6. 36   8. 1355   10. 1428   12. 35

Exercise Set 4.1.5

1. a. $7 \times 6 + 9 \times 2$
   $42 + 18$
   60

   b. $15 \times 2 + 12 \times 3$
   $30 + 36$
   66

   c. $9 \times 4 - 2 \times 7$
   $36 - 14$
   22

   d. $17 \times 4 - 18 \times 3$
   $68 - 54$
   14

   e. $4 \times 3 + 7 \times 3 - 4 \times 2$
   $12 + 21 - 8$
   25

   f. $12 \times 5 - 8 \times 4 + 14 \times 3$
   $60 - 32 + 42$
   70

   g. $9 + 4 \times 6$
   $9 + 24$
   33

   h. $12 - 4 \times 2$
   $12 - 8$
   4

   i. $14 \times 11 - 10$
   $154 - 10$
   144

   j. $27 + 5 \times 4 - 2$
   $27 + 20 - 2$
   45

   k. $19 \times 2 + 3 - 2 \times 8$
   $38 + 3 - 16$
   25

   l. $14 \times 3 \times 4 + 15 \times 3 \times 6$
   $168 + 270$
   438

   m. $9 \times 3 \times 2 - 5 \times 3$
   $54 - 15$
   39

   n. $3 \times 4 + 5 \times 4$
   $12 + 20$
   32

   o. $6 + 7 \times 2$
   $6 + 14$
   20

   p. $6 + 6 \times 2 - 1$
   $6 + 12 - 1$
   17

   q. $13 \, (2) + 14 \, (3)$
   $26 + 42$
   68

   r. $5 \, (4) \, (2) + 9 \, (3) \, (7)$
   $40 + 189$
   229

   s. $52 - 5 \, (3) \, (2)$
   $52 - 30$
   22

   t. $16 \, (2) \, (5) + 12$
   $160 + 12$
   172

   u. $17 \, (3) + 2 \, (3)$
   $51 + 6$
   57

2. a. $2 \, (3 + 2 \times 5) + 3 \times 2$
   $2 \, (3 + 10) + 6$
   $2 \times 13 + 6$
   $26 + 6$
   32

b. $5 \times 6 + 4 (12 - 2 \times 2)$
   $30 + 4 (12 - 4)$
   $30 + 4 \times 8$
   $30 + 32$
   62

c. $35 - 3 (4 \times 3 - 3 \times 2) + 4 (3)$
   $35 - 3 (12 - 6) + 12$
   $35 - 3 \times 6 + 12$
   $35 - 18 + 12$
   29

d. $2 (3 \times 5 + 2 \times 3) + 4 (24 - 2 \times 7)$
   $2 (15 + 6) + 4 (24 - 14)$
   $2 \times 21 + 4 \times 10$
   $42 + 40$
   82

e. $(3 + 4 \times 2) (2 \times 3 - 4) + 3 (9 - 4 \times 2)$
   $(3 + 8) (6 - 4) + 3 (9 - 8)$
   $11 \times 2 + 3 \times 1$
   $22 + 3$
   25

f. $(3 + 7) (9 - 2) (5 + 6)$
   $10 \times 7 \times 11$
   770

g. $3 (6 \times 2 + 2 \times 3) + 3 \times 8$
   $3 (12 + 6) + 24$
   $3 \times 18 + 24$
   $54 + 24$
   78

h. $3 (2 + 3 \times 7) (4 \times 2 - 1) + 4 (15 \times 3 + 4 \times 0)$
   $3 (2 + 21) (8 - 1) + 4 (45 + 0)$
   $3 \times 23 \times 7 + 4 \times 45$
   $483 + 180$
   663

i. $5 \times 6 + 2 \left[8 + 3 (4 - 2 \times 1) \right]$

   $30 + 2 \left[8 + 3 (4 - 2) \right]$

   $30 + 2 (8 + 3 \times 2)$
   $30 + 2 (8 + 6)$
   $30 + 2 \times 14$
   $30 + 28$
   58

j. $3 \left[2 (4 \times 5 - 2 \times 6) + 5 (3 \times 2 + 1) \right]$

   $3 \left[2 (20 - 12) + 5 (6 + 1) \right]$

   $3 (2 \times 8 + 5 \times 7)$

$3\,(16\,+\,35)$
3×51
153

k. $8\,(3\,+\,2\times6)\,-\,4\Big[15\,-\,(4\times5\,-\,6)\Big]$

$8\,(3\,+\,12)\,-\,4\Big[15\,-\,(20\,-\,6)\Big]$

$8\times15\,-\,4\,(15\,-\,14)$
$120\,-\,4\times1$
$120\,-\,4$
116

l. $3\,(2\,+\,6)\,-\,4\times5\,+\,8\Big[6\,-\,2\,(4\,-\,1)\Big]$
$3\times8\,-\,20\,+\,8\,(6\,-\,2\times3)$
$24\,-\,20\,+\,8\,(6\,-\,6)$
$24\,-\,20\,+\,8\times0$
$24\,-\,20\,+\,0$
4

m. $12\Big[30\,-\,4\,(5\,+\,2\times6)\,+\,8\,(19\,-\,2\times5)\Big]$

$12\Big[30\,-\,4\,(5\,+\,12)\,+\,8\,(19\,-\,10)\Big]$

$12\,(30\,-\,4\times17\,+\,8\times9)$
$12\,(30\,-\,68\,+\,72)$
12×34
408

n. $\Big[26\,-\,3\,(15\,-\,2\times6)\Big]\Big[4\,(3\times2\,-\,2\times1)\,+\,16\Big]$

$\Big[26\,-\,3\,(15\,-\,12)\Big]\Big[4\,(6\,-\,2)\,+\,16\Big]$

$(26\,-\,3\times3)\,(4\times4\,+\,16)$
$(26\,-\,9)\,(16\,+\,16)$
17×32
544

o. $2\Big[3\,(14\,-\,4\times3)\,-\,4\,(5\times6\,-\,3\times10)\Big]\,+\,3\,(4\times6\,-\,3\times2)$

$2\Big[3\,(14\,-\,12)\,-\,4\,(30\,-\,30)\Big]\,+\,3\,(24\,-\,6)$

$2\,(3\times2\,-\,4\times0)\,+\,3\times18$
$2\,(6\,-\,0)\,+\,54$
$2\times6\,+\,54$
$12\,+\,54$
66

3. a. $4\,+\,5\times2\,-\,10\div5$
$4\,+\,10\,-\,2$
12

b. $14\div2\times7\,+\,3$
$49\,+\,3$
52

c. $84 \div 2 \times 3 \div 7$
   18
d. $3 + 2 \times 5 + 7 \div 7 \times 2$
   $3 + 10 + 2$
   15
e. $3 + 6 \times 3 \times 2 - 1 + 4 \div 2$
   $3 + 36 - 1 + 2$
   40
f. $4 \div 2 + 5 \times 3 \div 15 + 4$
   $2 + 1 + 4$
   7
g. $2(3 \times 2 + 2 \times 4) + 3 \div 3$

$2(6 + 8) + 1$
$2 \times 14 + 1$
$28 + 1$
29

h. $100 + 20 \times 6 \div 2 - 100 \div 50 \times 2$
   $100 + 60 - 4$
   156

i. $4 \times 3 + 10 \div 5 \times 2 - 4(3 \times 2 - 6)$
   $12 + 4 - 4(6 - 6)$
   $12 + 4 - 4 \times 0$
   $12 + 4 - 0$
   16

Exercise Set 4.1.6

1. a. 1   b. 0   c. 40

2. a. 16   c. 22   e. 1
   b. 35   d. 570   f. 82

3. a.

   b.

   c.

4.

5.

6.

7.

8.

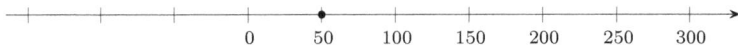

Exercise Set 4.2.1

1. 2
2. 47
3. 4100
4. 0
5. 32
6. 1

Exercise Set 4.2.2

1. negative two
2. positive six
3. negative one hundred ninety-eight
4. positive seven hundred eighty-two
5. negative one hundred four
6. the absolute value of negative seven
7. the absolute value of zero
8. the absolute value of twelve
9. negative fifteen
10. positive two hundred five
11. one thousand two hundred
12. the absolute value of positive four hundred ten
13. zero
14. negative one
15. negative twenty thousand four
16. the absolute value of one

Exercise Set 4.2.3

1. a. $-20 > -47$
   b. $-6271 < -5989$
   c. $-240 < 15$
   d. $-6 < 0$
   e. $|-25| > |1|$
   f. $|0| < |-20|$
   g. $|34| > |12|$
   h. $|18| < |-40|$

2. a. $0, -2, -4, -8, -16$
   b. $826, 552, -700, -950$
   c. $-744, -745, -746, -747$
   d. $-18, -84, -310, -420$

3. a. $-18, -3, 0, 2, 15$
   b. $-324, -150, -126, -120$
   c. $-1020, -425, -95, 240, 710$
   d. $-593, -592, -591, -590$

Exercise Set 4.2.4

1. a. 11   f. 6   k. 114   p. −2775   u. 15
  b. −1   g. −6   l. −604   q. 42   v. −15
  c. 1   h. −30   m. 2775   r. 42
  d. −11   i. 604   n. 2125   s. −42
  e. 30   j. −114 o. −2125   t. −42

2. a.

$$8 + (+6) = 8 + 6$$
$$= 14$$

g.

$$-8 + (-140) = -8 - 140$$
$$= -148$$

b.

$$20 - (-18) = 20 + 18$$
$$= 38$$

h.

$$-1 - (+10) = -1 - 10$$
$$= -11$$

c.

$$-3 + (+7) = -3 + 7$$
$$= 4$$

i.

$$23 + (+34) = 23 + 34$$
$$= 57$$

d.

$$-21 - (-8) = -21 + 8$$
$$= -13$$

j.

$$627 - (+123) = 627 - 123$$
$$= 504$$

e.

$$42 + (-12) = 42 - 12$$
$$= 30$$

k.

$$-195 + (-62) = -195 - 62$$
$$= -257$$

f.

$$31 - (+1) = 31 - 1$$
$$= 30$$

l.

$$425 + (-510) = 425 - 510$$
$$= -85$$

3. a. 70   d. 70   g. 4   j. 4
  b. −70   e. 0   h. −4   k. 0
  c. −70   f. 0   i. −4   l. 0

    This expression is undefined.
  m. This expression is undefined.
  o. This expression is indeterminate.

p. 5   q. -3 r. -50 s. -2 t. 2

u. This expression is undefined.

Exercise Set 4.2.5

1. a.
$$-9 \times 4 + 5(-6)(8) = -36 - 240$$
$$= -276$$

b.
$$-6 \times 4 \times 2 - (-7)(2) = -48 + 14$$
$$= -34$$

c.
$$-12 \div (-3) \div 2 + (-5)(-2) - 18 \div (-2)(-7)$$
$$= 2 + 10 - 63$$
$$= -51$$

d.
$$-2 \times 4 + 6(-8) = -8 - 48$$
$$= -56$$

e.
$$3(-2)(-4) - (-5)(6)$$
$$= 24 + 30$$
$$= 54$$

f.
$$-4 \times 2 \times 3 - 5(-3)(-8) = -24 - 120$$
$$= -144$$

g.
$$4(-7) + (-6)(-4) = -28 + 24$$
$$= -4$$

h.
$$-5(-8) - 3(-2) = 40 + 6$$
$$= 46$$

i.
$$4(-3) + 2(-1) - (-6) = -12 - 2 + 6$$
$$= -8$$

j.

$$3\Big[2 + 8(-1)\Big] - 2(-3) = 3(2 - 8) + 6$$
$$= 3(-6) + 6$$
$$= -18 + 6$$
$$= -12$$

k.

$$5 \times 4 + (-6)\Big[3 - 2(-2)\Big]$$
$$= 20 + (-6)(3 + 4)$$
$$= 20 + (-6)(7)$$
$$= 20 - 42$$
$$= -22$$

l.

$$-2\Big[4(-3) - (-2)(5)\Big] = -2(-12 + 10)$$
$$= -2(-2)$$
$$= 4$$

m.

$$-(2 + 3 \times 6) = -(2 + 18)$$
$$= -(20)$$
$$= -20$$

n.

$$-\Big[-8 - (-3)(-2)\Big] = -(-8 - 6)$$
$$= -(-14)$$
$$= 14$$

o.

$$-2(8 - 3) - (-4 + 2) = -2(5) - (-2)$$
$$= -10 + 2$$
$$= -8$$

Exercise Set 4.2.6

1. a. -24 c. -9 e. -430
   b. -65 d. 230 f. -21

2.

$$-50 \quad -40 \quad -30 \quad -20 \quad -10 \quad 0 \quad 10 \quad 20 \quad 30 \quad 40$$

3.

$$-50 \quad -40 \quad -30 \quad -20 \quad -10 \quad 0 \quad 10 \quad 20 \quad 30 \quad 40$$

4.

$$-25 \quad -20 \quad -15 \quad -10 \quad -5 \quad 0 \quad 5 \quad 10 \quad 15 \quad 20$$

5.

$$-25 \quad -20 \quad -15 \quad -10 \quad -5 \quad 0 \quad 5 \quad 10 \quad 15 \quad 20$$

6.

$$-500 \quad -400 \quad -300 \quad -200 \quad -100 \quad 0 \quad 100 \quad 200 \quad 300 \quad 400$$

7.

$$-500 \quad -400 \quad -300 \quad -200 \quad -100 \quad 0 \quad 100 \quad 200 \quad 300 \quad 400$$

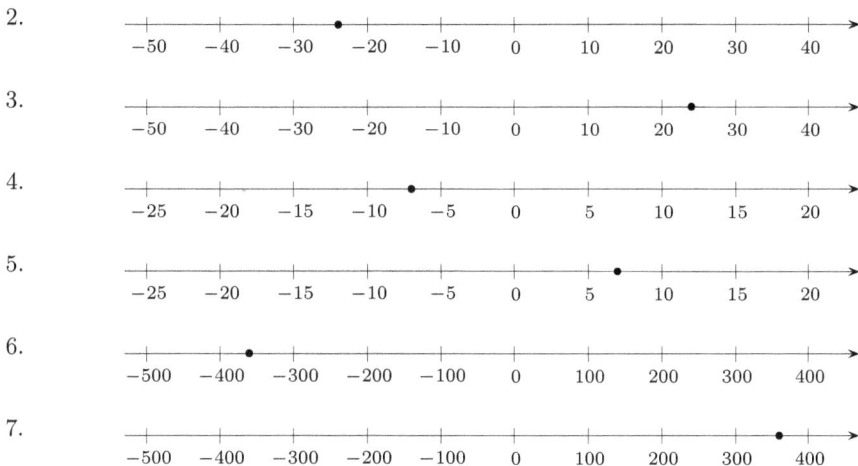

Exercise Set 4.3.0

1. 2, 3, 5, 7, 11, 13, 17, 19

2. a. 18 is divisible by 2 and 3.
   b. 25 is divisible by 5.
   c. 16 is divisible by 2.
   d. 21 is divisible by 3.
   e. 312 is divisible by 2 and 3.
   f. 420 is divisible by 2, 3, 5 and 10.
   g. 511 is not divisible by 2, 3, 5 or 10.
   h. 810 is divisible by 2, 3, 5 and 10.
   i. 4522 is divisible by 2.
   j. 343 is not divisible by 2, 3, 5 or 10.
   k. 1000 is divisible by 2, 5 and 10.
   l. 4 is divisible by 2.

Exercise Set 4.3.1

1. a. The number of equally-sized pieces that a whole should be cut into.
   b. The number of pieces of interest.

2. The size of a piece decreases. The size of a piece increases.

3. a. The fraction decreases in size. The fraction increases in size.
   b. The fraction increases in size. The fraction decreases in size.

4. a. $\frac{2}{5}$ b. $\frac{3}{5}$ c. $\frac{0}{10}$ d. $\frac{6}{6}$ e. $\frac{1}{7}$ f. $\frac{1}{1}$

5. a. improper fraction
    b. mixed number
    c. improper fraction
    d. improper fraction
    e. whole number
    f. proper fraction
    g. mixed number
    h. mixed number
    i. proper fraction
    j. none of the above
    k. improper fraction
    l. proper fraction
    m. none of the above
    n. proper fraction
    o. improper fraction

6. a. $6\frac{1}{2}$
    b. 2
    c. $1\frac{13}{15}$
    d. $6\frac{2}{3}$
    e. $2\frac{1}{4}$
    f. 34
    g. $62\frac{1}{18}$
    h. $52\frac{1}{100}$
    i. 2
    j. $3\frac{27}{100}$
    k. $12\frac{4}{25}$
    l. 8
    m. $6\frac{4}{9}$
    n. 1
    o. $1\frac{1}{24}$

7. a. $\frac{15}{2}$
    b. $\frac{152}{15}$
    c. $\frac{382}{365}$
    d. $\frac{97}{3}$
    e. $\frac{9}{4}$
    f. $\frac{182}{1}$
    g. $\frac{8}{1}$
    h. $\frac{17\,203}{4}$
    i. $\frac{1}{1}$
    j. $\frac{94}{5}$
    k. $\frac{17}{3}$
    l. $\frac{163}{8}$
    m. $\frac{12\,029}{120}$
    n. $\frac{481}{20}$
    o. $\frac{57}{8}$

8. a. $\frac{4}{5}$
    b. $\frac{2}{3}$
    c. $\frac{3}{11}$
    d. $\frac{4}{25}$
    e. $\frac{7}{20}$
    f. $\frac{11}{130}$
    g. $\frac{26}{317}$
    h. $\frac{17}{1500}$
    i. $\frac{8}{9}$
    j. $\frac{10}{49}$
    k. $\frac{2}{5}$
    l. $\frac{800}{863}$

9. a. $\frac{6}{11}$
    b. $\frac{4}{9}$
    c. $\frac{2}{7}$
    d. $\frac{11}{20}$
    e. $\frac{19}{23}$
    f. $\frac{15}{41}$
    g. $\frac{43}{191}$
    h. $\frac{8}{9}$
    i. $1\frac{3}{4}$
    j. $\frac{8}{9}$
    k. $\frac{15}{23}$
    l. $\frac{2}{3}$

10. a. $\frac{1}{3}$
    b. $\frac{2}{5}$
    c. $\frac{8}{25}$
    d. $\frac{16}{25}$
    e. $\frac{4}{5}$
    f. $\frac{3}{8}$
    g. $\frac{19}{173}$
    h. $\frac{1}{23}$
    i. $\frac{16}{17}$
    j. $\frac{15}{77}$
    k. $\frac{26}{31}$
    l. $\frac{1}{8}$

11. a. $\frac{5}{7}$
    b. $\frac{3}{20}$
    c. $\frac{37}{42}$
    d. $\frac{5}{56}$
    e. $\frac{2}{13}$
    f. $\frac{4}{7}$
    g. $\frac{14}{29}$
    h. $\frac{3}{13}$
    i. $\frac{6}{25}$
    j. $\frac{18}{35}$
    k. $\frac{32}{55}$
    l. $\frac{4}{9}$

12. a. $\frac{1}{2}$ e. $\frac{7}{9}$ i. $\frac{1}{7}$ m. $\frac{11}{20}$

   b. $\frac{5}{8}$ f. $\frac{1}{8}$ j. $15\frac{3}{4}$ n. $\frac{1}{6}$

   c. $\frac{4}{5}$ g. $\frac{3}{10}$ k. $\frac{2}{3}$ o. $\frac{3}{4}$

   d. $\frac{3}{4}$ h. $20\frac{1}{3}$ l. $\frac{7}{16}$ p. $\frac{2}{5}$

13. a. $\frac{3}{4}$ e. $\frac{5}{9}$ i. $\frac{2}{3}$ m. $\frac{2}{7}$

   b. $\frac{7}{9}$ f. $1\frac{3}{7}$ j. $\frac{7}{8}$ n. $\frac{7}{9}$

   c. $\frac{3}{4}$ g. $\frac{2}{5}$ k. $\frac{1}{4}$ o. $\frac{11}{16}$

   d. $\frac{2}{5}$ h. $120\frac{1}{4}$ l. $1\frac{1}{4}$ p. $\frac{2}{5}$

14. a. $\frac{1}{7}$ f. $\frac{2}{5}$ k. $\frac{11}{24}$ p. $\frac{5}{9}$

   b. $\frac{1}{2}$ g. $\frac{7}{15}$ l. $\frac{1}{3}$ q. $\frac{2}{3}$

   c. $\frac{1}{3}$ h. $\frac{3}{4}$ m. $2\frac{5}{6}$ r. $\frac{1}{40}$

   d. $10\frac{6}{7}$ i. $\frac{2}{5}$ n. $\frac{3}{4}$ s. $\frac{32}{125}$

   e. $\frac{5}{7}$ j. $7\frac{3}{8}$ o. $\frac{1}{3}$ t. $\frac{1}{10}$

15. a. $\frac{1}{3}$ e. $\frac{18}{25}$ i. $\frac{3}{8}$ m. $\frac{2}{3}$

   b. $\frac{2}{5}$ f. $4\frac{1}{4}$ j. $\frac{2}{7}$ n. $\frac{1}{5}$

   c. $\frac{8}{9}$ g. $\frac{9}{20}$ k. 4 o. $\frac{3}{149}$

   d. $3\frac{3}{4}$ h. $\frac{15}{56}$ l. $\frac{3}{4}$ p. $\frac{1}{2}$

Exercise Set 4.3.2

1. a. five sixths
   b. one tenth
   c. twenty-five hundredths
   d. three halves
   e. zero halves
   f. eighteen eighteenths
   g. twelve ones
   h. seven quarters
   i. one third
   j. two sixteenths
   k. nine sevenths
   l. twelve twelfths
   m. forty-six hundredths
   n. fifteen thousandths
   o. seven fifty-eighths
   p. negative five thirds
   q. negative eight fifteenths
   r. negative four fifths
   s. negative three thirds
   t. negative one twenty-fourth

2. a. $\frac{4}{5}$ e. $\frac{0}{100}$ i. $\frac{25}{10}$ m. $-\frac{8}{7}$ q. $-\frac{16}{16}$

   b. $\frac{3}{10}$ f. $\frac{1}{15}$ j. $+\frac{1}{1000}$ n. $\frac{6}{1}$ r. $\frac{1}{2}$

   c. $\frac{18}{3}$ g. $\frac{2}{2}$ k. $\frac{1}{6}$ o. $\frac{21}{39}$ s. $\frac{69}{4}$

   d. $\frac{42}{6}$ h. $\frac{78}{100}$ l. $\frac{9}{55}$ p. $+\frac{40}{7}$ t. $-\frac{80}{4}$

3. a. four and three fifths   e. one hundred twenty and one tenth
   b. three fifths   f. fifteen and one half
   c. fourteen thirds   g. eight and seven hundredths
   d. seven and two thirds   h. nine ten thousandths

4. a. $9\frac{3}{5}$ d. $32\frac{1}{8}$ g. $+5\frac{2}{3}$ j. $-2\frac{1}{5}$ m. $-1\frac{5}{7}$

   b. $2\frac{1}{2}$ e. $-14\frac{27}{100}$ h. $5\frac{2}{3}$ k. $17\frac{3}{10}$

   c. $15\frac{3}{4}$ f. $44\frac{18}{19}$ i. $-5\frac{2}{3}$ l. $+435\frac{79}{100}$

Exercise Set 4.3.3

1. a. $\frac{3}{4} > \frac{2}{4}$ e. $\frac{-3}{2} > \frac{5}{-2}$ i. $-\frac{8}{5} < \frac{-7}{5}$

   b. $-\frac{1}{5} > -\frac{4}{5}$ f. $\frac{32}{2} > \frac{12}{2}$ j. $\frac{20}{100} < \frac{40}{100}$

   c. $\frac{0}{3} < \frac{2}{3}$ g. $\frac{25}{50} < \frac{32}{50}$ k. $\frac{4}{4} > \frac{3}{4}$

   d. $\frac{520}{121} > \frac{519}{121}$ h. $\frac{6}{1} < \frac{8}{1}$ l. $\frac{50}{3} > \frac{18}{3}$

2. a. $\frac{3}{4} > \frac{2}{3}$ e. $\frac{-2}{2} < -\frac{3}{6}$ i. $\frac{8}{9} > \frac{3}{5}$

   b. $\frac{1}{5} < \frac{3}{4}$ f. $\frac{10}{4} < \frac{25}{7}$ j. $\frac{16}{5} < \frac{15}{2}$

   c. $-\frac{2}{5} > -\frac{3}{4}$ g. $\frac{-1}{2} > \frac{2}{-3}$ k. $-\frac{1}{4} > \frac{3}{-8}$

   d. $\frac{8}{20} = \frac{4}{10}$ h. $\frac{0}{4} < \frac{2}{5}$ l. $\frac{50}{3} < \frac{230}{12}$

3. a. $\frac{1}{5}, \frac{2}{5}, \frac{3}{4}$ e. $\frac{-5}{6}, -\frac{3}{4}, -\frac{2}{3}, -\frac{1}{5}$ i. $\frac{-7}{9}, \frac{-8}{15}, -\frac{2}{7}$

   b. $\frac{1}{4}, \frac{3}{7}, \frac{2}{3}$ f. $\frac{3}{15}, \frac{5}{11}, \frac{3}{2}$ j. $\frac{3}{8}, \frac{2}{5}, \frac{1}{2}$

   c. $\frac{1}{3}, \frac{1}{2}, \frac{3}{5}$ g. $-\frac{3}{4}, -\frac{4}{9}, -\frac{1}{4}$ k. $\frac{2}{3}, \frac{13}{15}, \frac{12}{11}, \frac{11}{10}$

   d. $-\frac{11}{20}, \frac{-8}{15}, \frac{7}{-15}, \frac{-2}{5}$ h. $\frac{1}{3}, \frac{4}{5}, \frac{4}{3}, \frac{7}{3}$ l. $\frac{4}{20}, \frac{2}{5}, \frac{3}{4}$

4. a. $3\frac{1}{4} > 2\frac{3}{5}$ b. $14\frac{1}{5} > 6\frac{3}{4}$ c. $-2\frac{3}{5} < -2\frac{4}{7}$

d. $6\frac{1}{2} < 6\frac{2}{3}$ g. $20\frac{1}{4} > 12\frac{4}{9}$ j. $6\frac{1}{20} < 6\frac{2}{25}$

e. $10\frac{14}{20} = 10\frac{7}{10}$ h. $-13\frac{2}{3} > -40\frac{2}{5}$ k. $-2\frac{1}{5} > -2\frac{3}{10}$

f. $1\frac{3}{15} < 1\frac{5}{20}$ i. $5\frac{7}{9} > 5\frac{2}{7}$ l. $7\frac{8}{10} = 7\frac{4}{5}$

5. a. $\frac{3}{10}, \frac{7}{10}, \frac{9}{10}$ i. $-\frac{9}{10}, -\frac{3}{5}, \frac{9}{20}, \frac{3}{4}$

b. $\frac{1}{8}, \frac{3}{8}, \frac{7}{8}$ j. $\frac{3}{25}, \frac{1}{5}, \frac{2}{5}, \frac{4}{5}$

c. $2\frac{1}{5}, 2\frac{3}{5}, 2\frac{4}{5}$ k. $1\frac{1}{2}, 2\frac{1}{5}, 3\frac{1}{4}, 4\frac{2}{3}$

d. $-\frac{3}{10}, -\frac{1}{4}, -\frac{1}{5}$ l. $-5\frac{2}{3}, -5\frac{1}{3}, 3\frac{2}{5}, 3\frac{2}{3}$

e. $\frac{2}{25}, \frac{1}{10}, \frac{3}{20}$ m. $18\frac{3}{100}, 18\frac{1}{20}, 18\frac{3}{20}, 18\frac{1}{4}$

f. $8\frac{1}{4}, 8\frac{1}{2}, 8\frac{3}{4}$ n. $9\frac{1}{5}, 9\frac{1}{2}, 9\frac{2}{3}, 9\frac{3}{4}, 9\frac{4}{5}$

g. $\frac{-19}{20}, \frac{9}{-10}, \frac{-4}{5}, -\frac{3}{4}$ o. $3\frac{1}{5}, 3\frac{1}{3}, 3\frac{2}{3}, 3\frac{3}{4}$

h. $\frac{2}{25}, \frac{9}{50}, \frac{1}{5}, \frac{3}{10}$ p. $2, 2\frac{1}{2}, 2\frac{2}{3}, 2\frac{3}{4}$

6. a. $\frac{14}{25}, \frac{9}{25}, \frac{3}{25}$ h. $8\frac{3}{4}, 8\frac{1}{2}, 8\frac{2}{5}$

b. $\frac{9}{100}, \frac{7}{100}, \frac{3}{100}$ i. $\frac{9}{10}, \frac{4}{5}, \frac{3}{4}, \frac{1}{2}$

c. $4\frac{11}{17}, 4\frac{8}{17}, 4\frac{3}{17}, 4\frac{1}{17}$ j. $92\frac{3}{4}, 92\frac{1}{2}, 92\frac{1}{3}, 91\frac{1}{2}$

d. $\frac{1}{2}, \frac{1}{3}, \frac{1}{5}, \frac{1}{10}$ k. $3\frac{1}{2}, \frac{4}{5}, -\frac{1}{5}, -\frac{3}{10}$

e. $\frac{1}{-2}, -\frac{3}{5}, \frac{-2}{3}$ l. $40\frac{3}{4}, 40\frac{2}{3}, 40\frac{1}{2}$

f. $\frac{4}{5}, \frac{2}{3}, \frac{1}{5}$ m. $3\frac{1}{2}, 3\frac{2}{5}, 2\frac{3}{10}, 2\frac{1}{5}$

g. $-3\frac{1}{6}, -3\frac{1}{3}, -3\frac{4}{5}$ n. $18\frac{1}{3}, -18, -18\frac{2}{5}$

Exercise Set 4.3.4

1. a.

$$\frac{2}{5} + \frac{1}{5} = \frac{2+1}{5}$$
$$= \frac{3}{5}$$

b.

$$\frac{1}{8} + \frac{2}{8} = \frac{1+2}{8}$$
$$= \frac{3}{8}$$

c.

$$\frac{3}{4} + \frac{2}{4} = \frac{3+2}{4}$$
$$= \frac{5}{4}$$
$$= 1\frac{1}{4}$$

d.

$$\frac{2}{3} + \frac{1}{3} = \frac{2 + 1}{3}$$
$$= \frac{3}{3}$$
$$= 1$$

e.

$$\frac{4}{9} + \frac{2}{9} = \frac{4 + 2}{9}$$
$$= \frac{6}{9}$$
$$= \frac{2}{3}$$

f.

$$\frac{3}{4} + \frac{3}{4} = \frac{3 + 3}{4}$$
$$= \frac{6}{4}$$
$$= \frac{3}{2}$$
$$= 1\frac{1}{2}$$

g.

$$\frac{1}{2} + \frac{1}{2} = \frac{1 + 1}{2}$$
$$= \frac{2}{2}$$
$$= 1$$

h.

$$\frac{1}{10} + \frac{7}{10} = \frac{1 + 7}{10}$$
$$= \frac{8}{10}$$
$$= \frac{4}{5}$$

i.

$$\frac{12}{18} + \frac{15}{18} = \frac{12 + 15}{18}$$
$$= \frac{27}{18}$$
$$= \frac{3}{2}$$
$$= 1\frac{1}{2}$$

j.

$$\frac{4}{10} + \frac{9}{10} = \frac{4 + 9}{10}$$
$$= \frac{13}{10}$$
$$= 1\frac{3}{10}$$

k.

$$\frac{1}{3} + \frac{1}{3} = \frac{1 + 1}{3}$$
$$= \frac{2}{3}$$

l.

$$\frac{0}{9} + \frac{3}{9} = \frac{0 + 3}{9}$$
$$= \frac{3}{9}$$
$$= \frac{1}{3}$$

2. a.

$$\frac{1}{3} + \frac{2}{7} = \frac{7 + 6}{21}$$
$$= \frac{13}{21}$$

b.

$$\frac{2}{5} + \frac{2}{10} = \frac{4 + 2}{10}$$
$$= \frac{6}{10}$$
$$= \frac{3}{5}$$

c.

$$\frac{3}{4} + \frac{1}{8} = \frac{6 + 1}{8}$$
$$= \frac{7}{8}$$

d.

$$\frac{4}{10} + \frac{7}{20} = \frac{8 + 7}{20}$$
$$= \frac{15}{20}$$
$$= \frac{3}{4}$$

e.

$$\frac{4}{7} + \frac{1}{9} = \frac{36 + 7}{63}$$
$$= \frac{43}{63}$$

f.

$$\frac{11}{50} + \frac{6}{25} = \frac{11 + 12}{50}$$
$$= \frac{23}{50}$$

g.

$$\frac{32}{40} + \frac{17}{30} = \frac{96 + 68}{120}$$
$$= \frac{164}{120}$$
$$= \frac{41}{30}$$
$$= 1\frac{11}{30}$$

h.

$$\frac{1}{25} + \frac{1}{20} = \frac{4 + 5}{100}$$
$$= \frac{9}{100}$$

i.

$$\frac{12}{33} + \frac{19}{22} = \frac{24 + 57}{66}$$
$$= \frac{81}{66}$$
$$= \frac{27}{22}$$
$$= 1\frac{5}{22}$$

j.

$$\frac{11}{34} + \frac{4}{51} = \frac{33 + 8}{102}$$
$$= \frac{41}{102}$$

k.

$$\frac{1}{15} + \frac{15}{32} = \frac{32 + 225}{480}$$
$$= \frac{257}{480}$$

l.

$$\frac{11}{19} + \frac{2}{17} = \frac{187 + 38}{323}$$
$$= \frac{225}{323}$$

3. a.

Alternative:

$$2\frac{1}{3} + 1\frac{3}{5} = \frac{7}{3} + \frac{8}{5}$$
$$= \frac{35 + 24}{15}$$
$$= \frac{59}{15}$$
$$= 3\frac{14}{15}$$

$$2\frac{1}{3} + 1\frac{3}{5} = 3\frac{5 + 9}{15}$$
$$= 3\frac{14}{15}$$

b.

$$8\frac{1}{4} + \frac{1}{2} = \frac{33}{4} + \frac{1}{2}$$
$$= \frac{33 + 2}{4}$$
$$= \frac{35}{4}$$
$$= 8\frac{3}{4}$$

Alternative:

$$8\frac{1}{4} + \frac{1}{2} = 8\frac{1 + 2}{4}$$
$$= 8\frac{3}{4}$$

c.

$$2\frac{3}{5} + 3 = \frac{13}{5} + \frac{3}{1}$$
$$= \frac{13 + 15}{5}$$
$$= \frac{28}{5}$$
$$= 5\frac{3}{5}$$

Alternative:

$$2\frac{3}{5} + 3 = 5\frac{3}{5}$$

d.

$$12 + 1\frac{3}{8} = \frac{12}{1} + \frac{11}{8}$$
$$= \frac{96 + 11}{8}$$
$$= \frac{107}{8}$$
$$= 13\frac{3}{8}$$

Alternative:

$$12 + 1\frac{3}{8} = 13\frac{3}{8}$$

e.

$$4 + \frac{2}{3} = \frac{4}{1} + \frac{2}{3}$$
$$= \frac{12 + 2}{3}$$
$$= \frac{14}{3}$$
$$= 4\frac{2}{3}$$

Alternative:

$$4 + \frac{2}{3} = 4\frac{2}{3}$$

f.

$$\frac{1}{9} + 2 = \frac{1}{9} + \frac{2}{1}$$
$$= \frac{1 + 18}{9}$$
$$= \frac{19}{9}$$
$$= 2\frac{1}{9}$$

Alternative:

$$\frac{1}{9} + 2 = 2\frac{1}{9}$$

g.

$$182\frac{3}{4} + \frac{425}{3} = \frac{731}{4} + \frac{425}{3}$$
$$= \frac{2193 + 1700}{12}$$
$$= \frac{3893}{12}$$
$$= 324\frac{5}{12}$$

Alternative:

$$182\frac{3}{4} + \frac{425}{3} = 182\frac{3}{4} + 141\frac{2}{3}$$
$$= 323\frac{9 + 8}{12}$$
$$= 323\frac{17}{12}$$
$$= 323 + \frac{17}{12}$$
$$= 323 + 1\frac{5}{12}$$
$$= 324\frac{5}{12}$$

h.

$$-\frac{3}{8} + \frac{2}{5} = \frac{-3}{8} + \frac{2}{5}$$
$$= \frac{-15 + 16}{40}$$
$$= \frac{1}{40}$$

i.

$$-4\frac{1}{5} + 2\frac{1}{2} = -\frac{21}{5} + \frac{5}{2}$$

$$= \frac{-21}{5} + \frac{5}{2}$$

$$= \frac{-42 + 25}{10}$$

$$= \frac{-17}{10}$$

$$= -\frac{17}{10}$$

$$= -1\frac{7}{10}$$

k.

$$42 + 57\frac{1}{3} = \frac{42}{1} + \frac{172}{3}$$

$$= \frac{126 + 172}{3}$$

$$= \frac{298}{3}$$

$$= 99\frac{1}{3}$$

Alternative:

$$42 + 57\frac{1}{3} = 99\frac{1}{3}$$

j.

$$-6 + 3\frac{2}{5} = \frac{-6}{1} + \frac{17}{5}$$

$$= \frac{-30 + 17}{5}$$

$$= \frac{-13}{5}$$

$$= -\frac{13}{5}$$

$$= -2\frac{3}{5}$$

l.

$$-2\frac{2}{5} + 7 = -\frac{12}{5} + \frac{7}{1}$$

$$= \frac{-12}{5} + \frac{7}{1}$$

$$= \frac{-12 + 35}{5}$$

$$= \frac{23}{5}$$

$$= 4\frac{3}{5}$$

4. a.

$$\frac{4}{5} - \frac{3}{5} = \frac{4-3}{5}$$

$$= \frac{1}{5}$$

b.

$$\frac{15}{22} - \frac{7}{22} = \frac{15-7}{22}$$

$$= \frac{8}{22}$$

$$= \frac{4}{11}$$

c.

$$\frac{3}{4} - \frac{1}{4} = \frac{3-1}{4}$$

$$= \frac{2}{4}$$

$$= \frac{1}{2}$$

d.

$$\frac{30}{25} - \frac{10}{25} = \frac{30-10}{25}$$

$$= \frac{20}{25}$$

$$= \frac{4}{5}$$

e.

$$\frac{20}{105} - \frac{14}{105} = \frac{20-14}{105}$$

$$= \frac{6}{105}$$

$$= \frac{2}{35}$$

f.

$$\frac{32}{5} - \frac{18}{5} = \frac{32-18}{5}$$

$$= \frac{14}{5}$$

$$= 2\frac{4}{5}$$

g.

$$\frac{19}{3} - \frac{8}{3} = \frac{19 - 8}{3}$$
$$= \frac{11}{3}$$
$$= 3\frac{2}{3}$$

j.

$$\frac{14}{5} - \frac{8}{5} = \frac{14 - 8}{5}$$
$$= \frac{6}{5}$$
$$= 1\frac{1}{5}$$

h.

$$\frac{3}{8} - \frac{1}{8} = \frac{3 - 1}{8}$$
$$= \frac{2}{8}$$
$$= \frac{1}{4}$$

k.

$$\frac{305}{6} - \frac{292}{6} = \frac{305 - 292}{6}$$
$$= \frac{13}{6}$$
$$= 2\frac{1}{6}$$

i.

$$\frac{9}{8} - \frac{3}{8} = \frac{9 - 3}{8}$$
$$= \frac{6}{8}$$
$$= \frac{3}{4}$$

l.

$$\frac{46}{9} - \frac{40}{9} = \frac{46 - 40}{9}$$
$$= \frac{6}{9}$$
$$= \frac{2}{3}$$

5. a.

$$\frac{3}{4} - \frac{1}{2} = \frac{3 - 2}{4}$$
$$= \frac{1}{4}$$

e.

$$\frac{3}{10} - \frac{1}{8} = \frac{12 - 5}{40}$$
$$= \frac{7}{40}$$

b.

$$\frac{9}{16} - \frac{2}{12} = \frac{27 - 8}{48}$$
$$= \frac{19}{48}$$

f.

$$\frac{9}{10} - \frac{8}{25} = \frac{45 - 16}{50}$$
$$= \frac{29}{50}$$

c.

$$\frac{3}{10} - \frac{4}{15} = \frac{9 - 8}{30}$$
$$= \frac{1}{30}$$

g.

$$\frac{8}{3} - \frac{4}{5} = \frac{40 - 12}{15}$$
$$= \frac{28}{15}$$
$$= 1\frac{13}{15}$$

d.

$$\frac{4}{5} - \frac{1}{2} = \frac{8 - 5}{10}$$
$$= \frac{3}{10}$$

h.

$$\frac{19}{15} - \frac{11}{20} = \frac{76 - 33}{60}$$
$$= \frac{43}{60}$$

i.

$$\frac{3}{10} - \frac{1}{20} = \frac{6-1}{20}$$
$$= \frac{5}{20}$$
$$= \frac{1}{4}$$

k.

$$\frac{42}{5} - \frac{18}{25} = \frac{210-18}{25}$$
$$= \frac{192}{25}$$
$$= 7\frac{17}{25}$$

j.

$$\frac{102}{7} - \frac{206}{15} = \frac{1530-1442}{105}$$
$$= \frac{88}{105}$$

l.

$$\frac{19}{3} - \frac{17}{4} = \frac{76-51}{12}$$
$$= \frac{25}{12}$$
$$= 2\frac{1}{12}$$

6. a.

$$4\frac{3}{4} - 1\frac{1}{3} = \frac{19}{4} - \frac{4}{3}$$
$$= \frac{57-16}{12}$$
$$= \frac{41}{12}$$
$$= 3\frac{5}{12}$$

Alternative:

$$4\frac{3}{4} - 1\frac{1}{3} = 3\frac{9-4}{12}$$
$$= 3\frac{5}{12}$$

b.

$$2\frac{4}{5} - \frac{3}{4} = \frac{14}{5} - \frac{3}{4}$$
$$= \frac{56-15}{20}$$
$$= \frac{41}{20}$$
$$= 2\frac{1}{20}$$

Alternative:

$$2\frac{4}{5} - \frac{3}{4} = 2\frac{16-15}{20}$$
$$= 2\frac{1}{20}$$

c.

$$16 - 5\frac{4}{5} = \frac{16}{1} - \frac{29}{5}$$
$$= \frac{80-29}{5}$$
$$= \frac{51}{5}$$
$$= 10\frac{1}{5}$$

Alternative:

$$16 - 5\frac{4}{5} = 15\frac{5}{5} - 5\frac{4}{5}$$
$$= 10\frac{5-4}{5}$$
$$= 10\frac{1}{5}$$

d.

$$21\frac{1}{5} - 14\frac{2}{3} = \frac{106}{5} - \frac{44}{3}$$
$$= \frac{318-220}{15}$$
$$= \frac{98}{5}$$
$$= 19\frac{3}{5}$$

e.

$$18\frac{1}{3} - 12 = \frac{55}{3} - \frac{12}{1}$$
$$= \frac{55 - 36}{3}$$
$$= \frac{19}{3}$$
$$= 6\frac{1}{3}$$

Alternative:

$$18\frac{1}{3} - 12 = 6\frac{1}{3}$$

f.

$$25 - 18\frac{3}{4} = \frac{25}{1} - \frac{75}{4}$$
$$= \frac{100 - 75}{4}$$
$$= \frac{25}{4}$$
$$= 6\frac{1}{4}$$

Alternative:

$$25 - 18\frac{3}{4} = 24\frac{4}{4} - 18\frac{3}{4}$$
$$= 6\frac{4 - 3}{4}$$
$$= 6\frac{1}{4}$$

g.

$$12\frac{1}{5} - 10 = \frac{61}{5} - \frac{10}{1}$$
$$= \frac{61 - 50}{5}$$
$$= \frac{11}{5}$$
$$= 2\frac{1}{5}$$

Alternative:

$$12\frac{1}{5} - 10 = 2\frac{1}{5}$$

h.

$$87\frac{5}{12} - 63\frac{7}{8} = \frac{1049}{12} - \frac{511}{8}$$
$$= \frac{2098 - 1533}{24}$$
$$= \frac{565}{24}$$
$$= 23\frac{13}{24}$$

i.

$$95\frac{1}{3} - 809\frac{2}{5} = \frac{286}{3} - \frac{4047}{5}$$
$$= \frac{1430 - 12\,141}{15}$$
$$= \frac{-10\,711}{15}$$
$$= -\frac{10\,711}{15}$$
$$= -714\frac{1}{15}$$

j.

$$\frac{3}{4} - \frac{5}{6} = \frac{9 - 10}{12}$$
$$= \frac{-1}{12}$$
$$= -\frac{1}{12}$$

k.

$$3 - 4\frac{1}{9} = \frac{3}{1} - \frac{37}{9}$$
$$= \frac{27 - 37}{9}$$
$$= \frac{-10}{9}$$
$$= -\frac{10}{9}$$
$$= -1\frac{1}{9}$$

l.

$$5\frac{3}{4} - 8\frac{1}{3} = \frac{23}{4} - \frac{25}{3}$$
$$= \frac{69 - 100}{12}$$
$$= \frac{-31}{12}$$
$$= -\frac{31}{12}$$
$$= -2\frac{7}{12}$$

m.

$$2\frac{9}{10} - \frac{25}{4} = \frac{29}{10} - \frac{25}{4}$$
$$= \frac{58 - 125}{20}$$
$$= \frac{-67}{20}$$
$$= -\frac{67}{20}$$
$$= -3\frac{7}{20}$$

n.

$$-\frac{43}{10} - \frac{27}{6} = \frac{-43}{10} - \frac{27}{6}$$
$$= \frac{-129 - 135}{30}$$
$$= \frac{-264}{30}$$
$$= -\frac{264}{30}$$
$$= -\frac{44}{5}$$
$$= -8\frac{4}{5}$$

o.

$$-3\frac{1}{5} - \frac{2}{7} = -\frac{16}{5} - \frac{2}{7}$$
$$= \frac{-16}{5} - \frac{2}{7}$$
$$= \frac{-112 - 10}{35}$$
$$= \frac{-122}{35}$$
$$= -\frac{122}{35}$$
$$= -3\frac{17}{35}$$

Alternative:

$$-3\frac{1}{5} - \frac{2}{7} = -3\frac{7 + 10}{35}$$
$$= -3\frac{17}{35}$$

p.

$$-9\frac{5}{8} - 23\frac{1}{6} = -\frac{77}{8} - \frac{139}{6}$$
$$= \frac{-77}{8} - \frac{139}{6}$$
$$= \frac{-231 - 556}{24}$$
$$= \frac{-787}{24}$$
$$= -\frac{787}{24}$$
$$= -32\frac{19}{24}$$

Alternative:

$$-9\frac{5}{8} - 23\frac{1}{6} = -32\frac{15 + 4}{24}$$
$$= -32\frac{19}{24}$$

q.

$$-2 - \frac{3}{8} = \frac{-2}{1} - \frac{3}{8}$$
$$= \frac{-16 - 3}{8}$$
$$= \frac{-19}{8}$$
$$= -\frac{19}{8}$$
$$= -2\frac{3}{8}$$

Alternative:

$$-2 - \frac{3}{8} = -2\frac{3}{8}$$

r.

$$-5 - 2\frac{1}{2} = \frac{-5}{1} - \frac{5}{2}$$
$$= \frac{-10 - 5}{2}$$
$$= \frac{-15}{2}$$
$$= -\frac{15}{2}$$
$$= -7\frac{1}{2}$$

Alternative:

$$-5 - 2\frac{1}{2} = -7\frac{1}{2}$$

Alternative:

$$-4\frac{2}{7} - \frac{6}{1} = -4\frac{2}{7} - 6$$
$$= -10\frac{2}{7}$$

s.

$$-4\frac{2}{7} - \frac{6}{1} = -\frac{30}{7} - \frac{6}{1}$$
$$= \frac{-30}{7} - \frac{6}{1}$$
$$= \frac{-30 - 42}{7}$$
$$= \frac{-72}{7}$$
$$= -\frac{72}{7}$$
$$= -10\frac{2}{7}$$

t.

$$5\frac{2}{3} - 8 = \frac{17}{3} - \frac{8}{1}$$
$$= \frac{17 - 24}{3}$$
$$= \frac{-7}{3}$$
$$= -\frac{7}{3}$$
$$= -2\frac{1}{3}$$

7. a.

$$\frac{2}{9} \times \frac{3}{4} = \frac{1}{6}$$

h.

$$\frac{12}{12} \times \frac{3}{4} = \frac{3}{4}$$

b.

$$\frac{3}{16} \times \frac{4}{3} = \frac{1}{4}$$

i.

$$\frac{1}{2}\left(\frac{2}{1}\right) = \frac{1}{1}$$
$$= 1$$

c.

$$\frac{30}{52} \times \frac{4}{3} = \frac{10}{13}$$

j.

$$\frac{14}{3}\left(\frac{2}{7}\right) = \frac{4}{3}$$
$$= 1\frac{1}{3}$$

d.

$$\frac{3}{2} \times \frac{2}{3} = \frac{1}{1}$$
$$= 1$$

e.

$$\frac{14}{25} \times \frac{25}{14} = \frac{1}{1}$$
$$= 1$$

k.

$$\frac{5}{18}\left(\frac{3}{20}\right) = \frac{1}{24}$$

f.

$$\frac{24}{18} \times \frac{12}{20} = \frac{4}{5}$$

l.

$$\frac{3}{4}\left(\frac{16}{20}\right) = \frac{3}{5}$$

g.

$$\frac{5}{3} \times \frac{9}{8} = \frac{15}{8}$$
$$= 1\frac{7}{8}$$

m.

$$\frac{32}{11}\left(\frac{22}{16}\right) = \frac{4}{1}$$
$$= 4$$

n.
$$\frac{13}{10}\left(\frac{1}{26}\right) = \frac{1}{20}$$

o.
$$\frac{5}{80}\left(\frac{30}{2}\right) = \frac{15}{16}$$

p.
$$\frac{7}{10}\left(\frac{5}{14}\right) = \frac{1}{4}$$

8. a.
$$3\frac{1}{2} \times \frac{1}{4} = \frac{7}{2} \times \frac{1}{4}$$
$$= \frac{7}{8}$$

b.
$$2\frac{1}{3} \times 1\frac{1}{7} = \frac{7}{3} \times \frac{8}{7}$$
$$= \frac{8}{3}$$
$$= 2\frac{2}{3}$$

c.
$$-4\frac{2}{3} \times 2\frac{1}{2} = -\frac{14}{3} \times \frac{5}{2}$$
$$= -\frac{35}{3}$$
$$= -11\frac{2}{3}$$

d.
$$-\frac{1}{5} \times \frac{5}{6} = -\frac{1}{6}$$

e.
$$-3\left(\frac{2}{15}\right) = -\frac{3}{1}\left(\frac{2}{15}\right)$$
$$= -\frac{2}{5}$$

f.
$$-1\frac{1}{3}\left(-\frac{1}{4}\right) = 1\frac{1}{3}\left(\frac{1}{4}\right)$$
$$= \frac{4}{3}\left(\frac{1}{4}\right)$$
$$= \frac{1}{3}$$

g.
$$2\left(\frac{-5}{4}\right) = -2\left(\frac{5}{4}\right)$$
$$= -\frac{2}{1}\left(\frac{5}{4}\right)$$
$$= -\frac{5}{2}$$
$$= -2\frac{1}{2}$$

h.
$$3\frac{1}{8}(-16) = -3\frac{1}{8}(16)$$
$$= -\frac{25}{8}\left(\frac{16}{1}\right)$$
$$= -\frac{50}{1}$$
$$= -50$$

i.
$$\frac{-14}{3}(-2) = \frac{14}{3}(2)$$
$$= \frac{14}{3}\left(\frac{2}{1}\right)$$
$$= \frac{28}{3}$$
$$= 9\frac{1}{3}$$

j.
$$\frac{7}{-12} \times \frac{6}{7} = -\frac{7}{12} \times \frac{6}{7}$$
$$= -\frac{1}{2}$$

k.

$$-2\frac{1}{2}(3) = -\frac{5}{2}\left(\frac{3}{1}\right)$$
$$= -\frac{15}{2}$$
$$= -7\frac{1}{2}$$

l.

$$\frac{-2}{-15}\left(\frac{-3}{11}\right) = -\frac{2}{15}\left(\frac{3}{11}\right)$$
$$= -\frac{2}{55}$$

m.

$$-\frac{11}{2} \times \frac{4}{-33} = \frac{11}{2} \times \frac{4}{33}$$
$$= \frac{2}{3}$$

n.

$$\frac{-34}{-15} \times \frac{-20}{14} = -\frac{34}{15} \times \frac{20}{14}$$
$$= -\frac{68}{21}$$
$$= -3\frac{5}{21}$$

o.

$$14 \times \frac{1}{-21} = -14 \times \frac{1}{21}$$
$$= -\frac{14}{1} \times \frac{1}{21}$$
$$= -\frac{2}{3}$$

p.

$$-5\frac{1}{3} \times \frac{-1}{-4} = -5\frac{1}{3} \times \frac{1}{4}$$
$$= -\frac{16}{3} \times \frac{1}{4}$$
$$= -\frac{4}{3}$$
$$= -1\frac{1}{3}$$

9. a.

$$\frac{4}{3} \div \frac{6}{7} = \frac{4}{3} \times \frac{7}{6}$$
$$= \frac{14}{9}$$
$$= 1\frac{5}{9}$$

b.

$$\frac{3}{10} \div \frac{3}{5} = \frac{3}{10} \times \frac{5}{3}$$
$$= \frac{1}{2}$$

c.

$$\frac{2}{3} \div \frac{8}{6} = \frac{2}{3} \times \frac{6}{8}$$
$$= \frac{1}{2}$$

d.

$$\frac{16}{21} \div \frac{8}{35} = \frac{16}{21} \times \frac{35}{8}$$
$$= \frac{10}{3}$$
$$= 3\frac{1}{3}$$

e.

$$\frac{12}{15} \div \frac{8}{25} = \frac{12}{15} \times \frac{25}{8}$$
$$= \frac{5}{2}$$
$$= 2\frac{1}{2}$$

f.

$$\frac{42}{13} \div \frac{21}{26} = \frac{42}{13} \times \frac{26}{21}$$
$$= \frac{4}{1}$$
$$= 4$$

g.

$$\frac{3}{4} \div \frac{3}{4} = \frac{3}{4} \times \frac{4}{3}$$
$$= \frac{1}{1}$$
$$= 1$$

l.

$$\frac{18}{15} \div \frac{6}{10} = \frac{18}{15} \times \frac{10}{6}$$
$$= \frac{2}{1}$$
$$= 2$$

h.

$$\frac{8}{3} \div \frac{16}{9} = \frac{8}{3} \times \frac{9}{16}$$
$$= \frac{3}{2}$$
$$= 1\frac{1}{2}$$

m.

$$\frac{6}{20} \div \frac{8}{10} = \frac{6}{20} \times \frac{10}{8}$$
$$= \frac{3}{8}$$

n.

$$\frac{14}{15} \div \frac{7}{20} = \frac{14}{15} \times \frac{20}{7}$$
$$= \frac{8}{3}$$
$$= 2\frac{2}{3}$$

i.

$$\frac{9}{10} \div \frac{27}{100} = \frac{9}{10} \times \frac{100}{27}$$
$$= \frac{10}{3}$$
$$= 3\frac{1}{3}$$

o.

$$\frac{16}{15} \div \frac{20}{105} = \frac{16}{15} \times \frac{105}{20}$$
$$= \frac{28}{5}$$
$$= 5\frac{3}{5}$$

j.

$$\frac{6}{5} \div \frac{12}{5} = \frac{6}{5} \times \frac{5}{12}$$
$$= \frac{1}{2}$$

p.

$$\frac{32}{15} \div \frac{7}{30} = \frac{32}{15} \times \frac{30}{7}$$
$$= \frac{64}{7}$$
$$= 9\frac{1}{7}$$

k.

$$\frac{3}{25} \div \frac{6}{100} = \frac{3}{25} \times \frac{100}{6}$$
$$= \frac{2}{1}$$
$$= 2$$

10. a.

$$\frac{-5}{14} \div \frac{1}{-7} = \frac{5}{14} \div \frac{1}{7}$$
$$= \frac{5}{14} \times \frac{7}{1}$$
$$= \frac{5}{2}$$
$$= 2\frac{1}{2}$$

b.

$$\frac{3}{-4} \div 1\frac{1}{8} = -\frac{3}{4} \div 1\frac{1}{8}$$
$$= -\frac{3}{4} \div \frac{9}{8}$$
$$= -\frac{3}{4} \times \frac{8}{9}$$
$$= -\frac{2}{3}$$

c.

$$-1\frac{3}{8} \div \frac{22}{3} = -\frac{11}{8} \div \frac{22}{3}$$
$$= -\frac{11}{8} \times \frac{3}{22}$$
$$= -\frac{3}{16}$$

d.

$$-1\frac{1}{2} \div \left(-\frac{5}{6}\right) = 1\frac{1}{2} \div \frac{5}{6}$$
$$= \frac{3}{2} \div \frac{5}{6}$$
$$= \frac{3}{2} \times \frac{6}{5}$$
$$= \frac{9}{5}$$
$$= 1\frac{4}{5}$$

e.

$$4\frac{2}{3} \div \left(-3\frac{1}{2}\right) = -4\frac{2}{3} \div 3\frac{1}{2}$$
$$= -\frac{14}{3} \div \frac{7}{2}$$
$$= -\frac{14}{3} \times \frac{2}{7}$$
$$= -\frac{4}{3}$$
$$= -1\frac{1}{3}$$

f.

$$-3\frac{1}{2} \div \frac{-1}{-8} = -3\frac{1}{2} \div \frac{1}{8}$$
$$= -\frac{7}{2} \div \frac{1}{8}$$
$$= -\frac{7}{2} \times \frac{8}{1}$$
$$= -\frac{28}{1}$$
$$= -28$$

g.

$$-7\frac{1}{2} \div \frac{-3}{10} = 7\frac{1}{2} \div \frac{3}{10}$$
$$= \frac{15}{2} \div \frac{3}{10}$$
$$= \frac{15}{2} \times \frac{10}{3}$$
$$= \frac{25}{1}$$
$$= 25$$

h.

$$\frac{14}{5} \div \left(-\frac{7}{10}\right) = -\frac{14}{5} \div \frac{7}{10}$$
$$= -\frac{14}{5} \times \frac{10}{7}$$
$$= -\frac{4}{1}$$
$$= -4$$

i.

$$\frac{4}{3} \div \left(-1\frac{2}{3}\right) = -\frac{4}{3} \div 1\frac{2}{3}$$
$$= -\frac{4}{3} \div \frac{5}{3}$$
$$= -\frac{4}{3} \times \frac{3}{5}$$
$$= -\frac{4}{5}$$

j.

$$14 \div \frac{7}{8} = \frac{14}{1} \div \frac{7}{8}$$
$$= \frac{14}{1} \times \frac{8}{7}$$
$$= \frac{16}{1}$$
$$= 16$$

k.

$$-2 \div \frac{4}{5} = -\frac{2}{1} \div \frac{4}{5}$$
$$= -\frac{2}{1} \times \frac{5}{4}$$
$$= -\frac{5}{2}$$
$$= -2\frac{1}{2}$$

l.

$$16 \div \left(-\frac{1}{2}\right) = -16 \div \frac{1}{2}$$
$$= -\frac{16}{1} \div \frac{1}{2}$$
$$= -\frac{16}{1} \times \frac{2}{1}$$
$$= -\frac{32}{1}$$
$$= -32$$

o.

$$-8\frac{1}{2} \div 4 = -\frac{17}{2} \div \frac{4}{1}$$
$$= -\frac{17}{2} \times \frac{1}{4}$$
$$= -\frac{17}{8}$$
$$= -2\frac{1}{8}$$

m.

$$-2\frac{1}{3} \div 14 = -\frac{7}{3} \div \frac{14}{1}$$
$$= -\frac{7}{3} \times \frac{1}{14}$$
$$= -\frac{1}{6}$$

p.

$$-\frac{18}{7} \div 6 = -\frac{18}{7} \div \frac{6}{1}$$
$$= -\frac{18}{7} \times \frac{1}{6}$$
$$= -\frac{3}{7}$$

n.

$$-\frac{18}{21} \div (-6) = \frac{18}{21} \div 6$$
$$= \frac{18}{21} \div \frac{6}{1}$$
$$= \frac{18}{21} \times \frac{1}{6}$$
$$= \frac{1}{7}$$

11. a. Numerator: $\frac{2}{3}$

Denominator: $\frac{4}{5}$

b. Numerator: 3

Denominator: $\frac{2}{5}$

c. Numerator: $\frac{4}{5}$

Denominator: 1

d. Numerator: $6\frac{1}{2}$

Denominator: $\frac{1}{2}$

12. a.

$$\frac{\frac{2}{3}}{\frac{4}{5}} = \frac{2 \times 5}{3 \times 4}$$
$$= \frac{5}{6}$$

b.

$$\frac{\frac{9}{5}}{\frac{7}{15}} = \frac{9 \times 15}{5 \times 7}$$
$$= \frac{27}{7}$$
$$= 3\frac{6}{7}$$

c.

$$\frac{\frac{3}{8}}{\frac{3}{4}} = \frac{3 \times 4}{8 \times 3}$$

$$\cdot \quad = \frac{1}{2}$$

h.

$$\frac{\frac{2}{3}}{\frac{8}{-9}} = -\frac{\frac{2}{3}}{\frac{8}{9}}$$

$$= -\frac{2 \times 9}{3 \times 8}$$

$$= -\frac{3}{4}$$

d.

$$\frac{\frac{1}{2}}{\frac{10}{3}} = \frac{1 \times 3}{2 \times 10}$$

$$= \frac{3}{20}$$

i.

$$\frac{\frac{-2}{3}}{-\frac{5}{-8}} = -\frac{\frac{2}{3}}{\frac{5}{8}}$$

$$= -\frac{2 \times 8}{3 \times 5}$$

$$= -\frac{16}{15}$$

$$= -1\frac{1}{15}$$

e.

$$\frac{\frac{4}{15}}{\frac{3}{20}} = \frac{4 \times 20}{15 \times 3}$$

$$= \frac{16}{9}$$

$$= 1\frac{7}{9}$$

j.

$$\frac{\frac{4}{-5}}{\frac{-3}{4}} = \frac{\frac{4}{5}}{\frac{3}{4}}$$

$$= \frac{4 \times 4}{5 \times 3}$$

$$= \frac{16}{15}$$

$$= 1\frac{1}{15}$$

f.

$$\frac{\frac{14}{35}}{\frac{9}{5}} = \frac{14 \times 5}{35 \times 9}$$

$$= \frac{2}{9}$$

g.

$$\frac{\frac{-18}{15}}{\frac{27}{25}} = -\frac{\frac{18}{15}}{\frac{27}{25}}$$

$$= -\frac{18 \times 25}{15 \times 27}$$

$$= -\frac{10}{9}$$

$$= -1\frac{1}{9}$$

k.

$$\frac{\frac{1}{23}}{\frac{3}{46}} = \frac{1 \times 46}{23 \times 3}$$

$$= \frac{2}{3}$$

l.

n.

$$\frac{-4}{5\frac{1}{2}} = -\frac{4}{5\frac{1}{2}}$$

$$= -\frac{\frac{4}{1}}{\frac{11}{2}}$$

$$= -\frac{4 \times 2}{1 \times 11}$$

$$= -\frac{8}{11}$$

$$\frac{-3}{-\frac{2}{5}} = \frac{3}{\frac{2}{5}}$$

$$= \frac{\frac{3}{1}}{\frac{2}{5}}$$

$$= \frac{3 \times 5}{1 \times 2}$$

o.

$$= \frac{15}{2}$$

$$= 7\frac{1}{2}$$

$$\frac{4\frac{2}{3}}{3\frac{1}{4}} = \frac{\frac{14}{3}}{\frac{13}{4}}$$

$$= \frac{14 \times 4}{3 \times 13}$$

m.

$$= \frac{56}{39}$$

$$= 1\frac{17}{39}$$

p.

$$\frac{1}{\frac{2}{3}} = \frac{\frac{1}{1}}{\frac{2}{3}}$$

$$= \frac{1 \times 3}{2 \times 1}$$

$$= \frac{3}{2}$$

$$= 1\frac{1}{2}$$

$$\frac{3\frac{1}{4}}{2} = \frac{\frac{13}{4}}{\frac{2}{1}}$$

$$= \frac{13 \times 1}{4 \times 2}$$

$$= \frac{13}{8}$$

$$= 1\frac{5}{8}$$

Exercise Set 4.3.5

1. a.

$$\frac{2}{5} + \frac{5}{6} + \frac{1}{3} = \frac{12 + 25 + 10}{30}$$

$$= \frac{47}{30}$$

$$= 1\frac{17}{30}$$

b.

$$\frac{1}{4} + \frac{2}{5} + \frac{2}{3} = \frac{15 + 24 + 40}{60}$$
$$= \frac{79}{60}$$
$$= 1\frac{19}{60}$$

c.

$$\frac{3}{8} + \frac{1}{4} - \frac{1}{2} = \frac{3 + 2 - 4}{8}$$
$$= \frac{1}{8}$$

d.

$$\frac{5}{12} - \frac{1}{8} - \frac{1}{6} = \frac{10 - 3 - 4}{24}$$
$$= \frac{3}{24}$$
$$= \frac{1}{8}$$

e.

$$\frac{1}{12} + \frac{2}{3} - \frac{1}{5} + \frac{5}{6} = \frac{5 + 40 - 12 + 50}{60}$$
$$= \frac{83}{60}$$
$$= 1\frac{23}{60}$$

f.

$$-\frac{1}{15} - \frac{5}{12} + \frac{5}{8} = \frac{-1}{15} - \frac{5}{12} + \frac{5}{8}$$
$$= \frac{-8 - 50 + 75}{120}$$
$$= \frac{17}{120}$$

g.

$$-\frac{1}{17} + \frac{5}{34} - \frac{3}{2} = \frac{-1}{17} + \frac{5}{34} - \frac{3}{2}$$
$$= \frac{-2 + 5 - 51}{34}$$
$$= \frac{-48}{34}$$
$$= -\frac{24}{17}$$
$$= -1\frac{7}{17}$$

h.

$$\frac{3}{8} + \frac{1}{2} - \frac{1}{4} = \frac{3 + 4 - 2}{8}$$
$$= \frac{5}{8}$$

i.

$$-\frac{9}{14} + \frac{2}{21} - \frac{3}{7} = \frac{-9}{14} + \frac{2}{21} - \frac{3}{7}$$
$$= \frac{-27 + 4 - 18}{42}$$
$$= \frac{-41}{42}$$
$$= -\frac{41}{42}$$

j.

$$\frac{3}{16} - \frac{1}{4} + \frac{2}{3} = \frac{9 - 12 + 32}{48}$$
$$= \frac{29}{48}$$

k.

$$\frac{4}{25} + \frac{1}{15} + \frac{2}{3} = \frac{12 + 5 + 50}{75}$$
$$= \frac{67}{75}$$

l.

$$-\frac{7}{10} - \frac{1}{2} - \frac{3}{4} - \frac{1}{8} = \frac{-7}{10} - \frac{1}{2} - \frac{3}{4} - \frac{1}{8}$$
$$= \frac{-28 - 20 - 30 - 5}{40}$$
$$= \frac{-83}{40}$$
$$= -2\frac{3}{40}$$

m.

$$\frac{1}{3} + \frac{3}{4} - \frac{1}{4} = \frac{4 + 9 - 3}{12}$$
$$= \frac{10}{12}$$
$$= \frac{5}{6}$$

n.

$$\frac{11}{10} + \frac{1}{2} + \frac{1}{4} = \frac{22 + 10 + 5}{20}$$
$$= \frac{37}{20}$$
$$= 1\frac{17}{20}$$

o.

$$-\frac{2}{3} - \frac{2}{5} + \frac{15}{4} = \frac{-2}{3} - \frac{2}{5} + \frac{15}{4}$$
$$= \frac{-40 - 24 + 225}{60}$$
$$= \frac{161}{60}$$
$$= 2\frac{41}{60}$$

p.

$$\frac{3}{9} + \frac{1}{6} + \frac{4}{3} + \frac{2}{3} = \frac{6 + 3 + 24 + 12}{18}$$
$$= \frac{45}{18}$$
$$= \frac{5}{2}$$
$$= 2\frac{1}{2}$$

2. a.

$$\frac{4}{5} + 5\frac{1}{2} - 1\frac{3}{4} = \frac{4}{5} + \frac{11}{2} - \frac{7}{4}$$
$$= \frac{16 + 110 - 35}{20}$$
$$= \frac{91}{20}$$
$$= 4\frac{11}{20}$$

Alternative:

$$\frac{4}{5} + 5\frac{1}{2} - 1\frac{3}{4} = 4\frac{16 + 10 - 15}{20}$$
$$= 4\frac{11}{20}$$

b.

$$\frac{9}{5} - 1\frac{3}{4} - 2\frac{1}{5} = \frac{9}{5} - \frac{7}{4} - \frac{11}{5}$$
$$= \frac{36 - 35 - 44}{20}$$
$$= \frac{-43}{20}$$
$$= -2\frac{3}{20}$$

c.

$$4\frac{2}{3} + 1\frac{1}{2} - \frac{3}{4} = \frac{14}{3} + \frac{3}{2} - \frac{3}{4}$$
$$= \frac{56 + 18 - 9}{12}$$
$$= \frac{65}{12}$$
$$= 5\frac{5}{12}$$

Alternative:

$$4\frac{2}{3} + 1\frac{1}{2} - \frac{3}{4} = 5\frac{8 + 6 - 9}{12}$$
$$= 5\frac{5}{12}$$

d.

$$6\frac{2}{3} - 1 + 3\frac{1}{2} - 2\frac{1}{4} = \frac{20}{3} - \frac{1}{1} + \frac{7}{2} - \frac{9}{4}$$
$$= \frac{80 - 12 + 42 - 27}{12}$$
$$= \frac{83}{12}$$
$$= 6\frac{11}{12}$$

Alternative:

$$6\frac{2}{3} - 1 + 3\frac{1}{2} - 2\frac{1}{4} = 6\frac{8 + 6 - 3}{12}$$
$$= 6\frac{11}{12}$$

e.

$$\frac{3}{4} - 2\frac{1}{2} + 4 = \frac{3}{4} - \frac{5}{2} + \frac{4}{1}$$
$$= \frac{3 - 10 + 16}{4}$$
$$= \frac{9}{4}$$
$$= 2\frac{1}{4}$$

Alternative:

$$\frac{3}{4} - 2\frac{1}{2} + 4 = 2\frac{3 - 2}{4}$$
$$= 2\frac{1}{4}$$

f.

$$14\frac{2}{3} + 1\frac{1}{3} - 3\frac{2}{5} = \frac{44}{3} + \frac{4}{3} - \frac{17}{5}$$
$$= \frac{220 + 20 - 51}{15}$$
$$= \frac{189}{15}$$
$$= \frac{63}{5}$$
$$= 12\frac{3}{5}$$

Alternative:

$$14\frac{2}{3} + 1\frac{1}{3} - 3\frac{2}{5} = 12\frac{10 + 5 - 6}{15}$$
$$= 12\frac{9}{15}$$
$$= 12\frac{3}{5}$$

g.

$$15\frac{2}{3} + 6\frac{1}{2} - 12\frac{3}{4} = \frac{47}{3} + \frac{13}{2} - \frac{51}{4}$$
$$= \frac{188 + 78 - 153}{12}$$
$$= \frac{113}{12}$$
$$= 9\frac{5}{12}$$

Alternative:

$$15\frac{2}{3} + 6\frac{1}{2} - 12\frac{3}{4} = 9\frac{8 + 6 - 9}{12}$$
$$= 9\frac{5}{12}$$

h.

$$4\frac{3}{7} - 2\frac{1}{14} + 1 - 3\frac{1}{4} = \frac{31}{7} - \frac{29}{14} + \frac{1}{1} - \frac{13}{4}$$
$$= \frac{124 - 58 + 28 - 91}{28}$$
$$= \frac{3}{28}$$

Alternative:

$$4\frac{3}{7} - 2\frac{1}{14} + 1 - 3\frac{1}{4} = \frac{12 - 2 - 7}{28}$$
$$= \frac{3}{28}$$

i.

$$2\frac{1}{5} + 3\frac{3}{4} - 1\frac{1}{2} + 4\frac{2}{3} = \frac{11}{5} + \frac{15}{4} - \frac{3}{2} + \frac{14}{3}$$

$$= \frac{132 + 225 - 90 + 280}{60}$$

$$= \frac{547}{60}$$

$$= 9\frac{7}{60}$$

Alternative:

$$2\frac{1}{5} + 3\frac{3}{4} - 1\frac{1}{2} + 4\frac{2}{3} = 8\frac{12 + 45 - 30 + 40}{60}$$

$$= 8\frac{67}{60}$$

$$= 8 + \frac{67}{60}$$

$$= 8 + 1\frac{7}{60}$$

$$= 9\frac{7}{60}$$

j.

$$-3\frac{2}{5} + 1\frac{1}{2} - 2\frac{1}{4} = -\frac{17}{5} + \frac{3}{2} - \frac{9}{2}$$

$$= \frac{-17}{5} + \frac{3}{2} - \frac{9}{2}$$

$$= \frac{-34 + 15 - 45}{10}$$

$$= \frac{-64}{10}$$

$$= -\frac{32}{5}$$

$$= -6\frac{2}{5}$$

k.

$$14\frac{3}{4} - 11\frac{1}{2} + 2\frac{3}{5} = \frac{59}{4} - \frac{23}{2} + \frac{13}{5}$$

$$= \frac{295 - 230 + 52}{20}$$

$$= \frac{117}{20}$$

$$= 5\frac{17}{20}$$

Alternative:

$$14\frac{3}{4} - 11\frac{1}{2} + 2\frac{3}{5} = 5\frac{15 - 10 + 12}{20}$$

$$= 5\frac{17}{20}$$

l.

$$-6\frac{1}{3} - 2\frac{3}{4} - 1\frac{1}{3} = -\frac{19}{3} - \frac{11}{4} - \frac{4}{3}$$
$$= \frac{-19}{3} - \frac{11}{4} - \frac{4}{3}$$
$$= \frac{-76 - 33 - 16}{12}$$
$$= \frac{-125}{12}$$
$$= -10\frac{5}{12}$$

Alternative:

$$-6\frac{1}{3} - 2\frac{3}{4} - 1\frac{1}{3} = -9\frac{4 + 9 + 4}{12}$$
$$= -9\frac{17}{12}$$
$$= -9 - \frac{17}{12}$$
$$= -9 - 1\frac{5}{12}$$
$$= -10\frac{5}{12}$$

m.

$$12 - 10\frac{3}{5} + 8\frac{1}{2} = \frac{12}{1} - \frac{53}{5} + \frac{17}{2}$$
$$= \frac{120 - 106 + 85}{10}$$
$$= \frac{99}{10}$$
$$= 9\frac{9}{10}$$

n.

$$-160\frac{2}{3} + 200 + 360\frac{1}{3} = -\frac{482}{3} + \frac{200}{1} + \frac{1081}{3}$$
$$= \frac{-482}{3} + \frac{200}{1} + \frac{1081}{3}$$
$$= \frac{-482 + 600 + 1081}{3}$$
$$= \frac{1199}{3}$$
$$= 399\frac{2}{3}$$

o.

$$-2\frac{1}{3} - 4\frac{2}{5} + 5 + 3\frac{2}{3} = -\frac{7}{3} - \frac{22}{5} + \frac{5}{1} + \frac{11}{3}$$

$$= \frac{-7}{3} - \frac{22}{5} + \frac{5}{1} + \frac{11}{3}$$

$$= \frac{-35 - 66 + 75 + 55}{15}$$

$$= \frac{29}{15}$$

$$= 1\frac{14}{15}$$

p.

$$8\frac{1}{5} + 2\frac{1}{3} + 3\frac{1}{4} - 5\frac{1}{2} = \frac{41}{5} + \frac{7}{3} + \frac{13}{4} - \frac{11}{2}$$

$$= \frac{492 + 140 + 195 - 330}{60}$$

$$= \frac{497}{60}$$

$$= 8\frac{17}{60}$$

Alternative:

$$8\frac{1}{5} + 2\frac{1}{3} + 3\frac{1}{4} - 5\frac{1}{2} = 8\frac{12 + 20 + 15 - 30}{60}$$

$$= 8\frac{17}{60}$$

3. a.

$$\frac{2}{3} \times \frac{1}{2} \times \frac{3}{4} = \frac{1}{4}$$

b.

$$\frac{4}{5} \times 1\frac{1}{2} \times 15 = \frac{4}{5} \times \frac{3}{2} \times \frac{15}{1}$$

$$= \frac{18}{1}$$

$$= 18$$

c.

$$-2\frac{1}{3} \times 9 \times 4\frac{1}{2} = -\frac{7}{3} \times \frac{9}{1} \times \frac{9}{2}$$

$$= -\frac{189}{2}$$

$$= -94\frac{1}{2}$$

d.

$$
\begin{aligned}
4 \times 3\frac{1}{4} \times \frac{5}{26} &= \frac{4}{1} \times \frac{13}{4} \times \frac{5}{26} \\
&= \frac{5}{2} \\
&= 2\frac{1}{2}
\end{aligned}
$$

e.

$$
\begin{aligned}
16 \times \frac{-1}{144} \times 12 &= -\frac{16}{1} \times \frac{1}{144} \times \frac{12}{1} \\
&= -\frac{4}{3} \\
&= -1\frac{1}{3}
\end{aligned}
$$

f.

$$
\begin{aligned}
-4\frac{2}{3} \times \frac{1}{7} \times 24 \times 3\frac{1}{6} &= -\frac{14}{3} \times \frac{1}{7} \times \frac{24}{1} \times \frac{19}{6} \\
&= -\frac{152}{3} \\
&= -50\frac{2}{3}
\end{aligned}
$$

g.

$$
\begin{aligned}
9 \times \frac{4}{3} \times 2\frac{1}{3} \times \frac{3}{4} &= \frac{9}{1} \times \frac{4}{3} \times \frac{7}{3} \times \frac{3}{4} \\
&= \frac{21}{1} \\
&= 21
\end{aligned}
$$

h.

$$
\begin{aligned}
3\frac{4}{5} \times \frac{18}{19} \times 2\frac{1}{6} &= \frac{19}{5} \times \frac{18}{19} \times \frac{13}{6} \\
&= \frac{39}{5} \\
&= 7\frac{4}{5}
\end{aligned}
$$

i.

$$
\begin{aligned}
1\frac{5}{6} \times 1\frac{1}{2} \times 8 \times \frac{-4}{33} &= -\frac{11}{6} \times \frac{3}{2} \times \frac{8}{1} \times \frac{4}{33} \\
&= -\frac{8}{3} \\
&= -2\frac{2}{3}
\end{aligned}
$$

j.

$$-9 \times \frac{-2}{-3} \times 3\frac{1}{4} = -\frac{9}{1} \times \frac{2}{3} \times \frac{13}{4}$$
$$= -\frac{39}{2}$$
$$= -19\frac{1}{2}$$

k.

$$5\frac{1}{2} \times 34 \times \frac{3}{17} = \frac{11}{2} \times \frac{34}{1} \times \frac{3}{17}$$
$$= \frac{33}{1}$$
$$= 33$$

l.

$$1\frac{2}{15} \times \frac{2}{7} \times \frac{14}{34} = \frac{17}{15} \times \frac{2}{7} \times \frac{14}{34}$$
$$= \frac{2}{15}$$

m.

$$-32 \times 4\frac{1}{8} \times 2 \times \frac{-1}{-11} = -\frac{32}{1} \times \frac{33}{8} \times \frac{2}{1} \times \frac{1}{11}$$
$$= -\frac{24}{1}$$
$$= -24$$

n.

$$\frac{3}{4} \times 14 \times \frac{1}{21} \times 5 \times \frac{1}{15} = \frac{3}{4} \times \frac{14}{1} \times \frac{1}{21} \times \frac{5}{1} \times \frac{1}{15}$$
$$= \frac{1}{6}$$

4. a.

$$\frac{3}{14} \div \frac{5}{7} \times \frac{2}{9} = \frac{3}{14} \times \frac{7}{5} \times \frac{2}{9}$$
$$= \frac{1}{15}$$

b.

$$\frac{4}{5} \div 1\frac{1}{6} \times 10\frac{1}{2} = \frac{4}{5} \div \frac{7}{6} \times \frac{21}{2}$$
$$= \frac{4}{5} \times \frac{6}{7} \times \frac{21}{2}$$
$$= \frac{36}{5}$$
$$= 7\frac{1}{5}$$

c.

$$\frac{3}{4} \div \frac{3}{2} \div \frac{5}{4} \times \frac{1}{3} = \frac{3}{4} \times \frac{2}{3} \times \frac{4}{5} \times \frac{1}{3}$$
$$= \frac{2}{15}$$

d.

$$4\frac{4}{5} \div \frac{14}{15} \div \frac{6}{21} = \frac{24}{5} \div \frac{14}{15} \div \frac{6}{21}$$
$$= \frac{24}{5} \times \frac{15}{14} \times \frac{21}{6}$$
$$= \frac{18}{1}$$
$$= 18$$

e.

$$\frac{1}{4} \div \frac{2}{5} \times 5 = \frac{1}{4} \div \frac{2}{5} \times \frac{5}{1}$$
$$= \frac{1}{4} \times \frac{5}{2} \times \frac{5}{1}$$
$$= \frac{25}{8}$$
$$= 3\frac{1}{8}$$

f.

$$2\frac{2}{3} \div 3\frac{1}{2} \times 2\frac{4}{5} \times \frac{3}{4} = \frac{8}{3} \div \frac{7}{2} \times \frac{14}{5} \times \frac{3}{4}$$
$$= \frac{8}{3} \times \frac{2}{7} \times \frac{14}{5} \times \frac{3}{4}$$
$$= \frac{8}{5}$$
$$= 1\frac{3}{5}$$

g.

$$4\frac{2}{3} \div \frac{1}{2} \div \frac{1}{3} = \frac{14}{3} \div \frac{1}{2} \div \frac{1}{3}$$
$$= \frac{14}{3} \times \frac{2}{1} \times \frac{3}{1}$$
$$= \frac{28}{1}$$
$$= 28$$

h.

$$-16 \times 1\frac{1}{2} \div 3 = -\frac{16}{1} \times \frac{3}{2} \div \frac{3}{1}$$
$$= -\frac{16}{1} \times \frac{3}{2} \times \frac{1}{3}$$
$$= -\frac{8}{1}$$
$$= -8$$

i.

$$\frac{-1}{3} \times \frac{-1}{-4} \div \frac{-2}{5} \div \frac{1}{2} = \frac{1}{3} \times \frac{1}{4} \times \frac{5}{2} \times \frac{2}{1}$$
$$= \frac{5}{12}$$

j.

$$\frac{4}{3} \div 4\frac{2}{3} \times 1\frac{1}{5} \div \frac{2}{7} = \frac{4}{3} \div \frac{14}{3} \times \frac{6}{5} \div \frac{2}{7}$$
$$= \frac{4}{3} \times \frac{3}{14} \times \frac{6}{5} \times \frac{7}{2}$$
$$= \frac{6}{5}$$
$$= 1\frac{1}{5}$$

k.

$$-1\frac{1}{15} \div \frac{-7}{21} \times \frac{5}{-13} = -\frac{16}{15} \div \frac{7}{21} \times \frac{5}{13}$$
$$= -\frac{16}{15} \times \frac{21}{7} \times \frac{5}{13}$$
$$= -\frac{16}{13}$$
$$= -1\frac{3}{13}$$

l.

$$1\frac{2}{3} \times 1 \div \frac{-3}{4} \div \frac{1}{-2} = \frac{5}{3} \times \frac{1}{1} \div \frac{3}{4} \div \frac{1}{2}$$
$$= \frac{5}{3} \times \frac{1}{1} \times \frac{4}{3} \times \frac{2}{1}$$
$$= \frac{40}{9}$$
$$= 4\frac{4}{9}$$

m.

$$
\begin{aligned}
-36 \div \frac{-18}{-21} \times \frac{1}{-7} \times 2 &= \frac{36}{1} \div \frac{18}{21} \times \frac{1}{7} \times \frac{2}{1} \\
&= \frac{36}{1} \times \frac{21}{18} \times \frac{1}{7} \times \frac{2}{1} \\
&= \frac{12}{1} \\
&= 12
\end{aligned}
$$

n.

$$
\begin{aligned}
8\frac{1}{2} \div 4 \div 3 &= \frac{17}{2} \div \frac{4}{1} \div \frac{3}{1} \\
&= \frac{17}{2} \times \frac{1}{4} \times \frac{1}{3} \\
&= \frac{17}{24}
\end{aligned}
$$

o.

$$
\begin{aligned}
\frac{3}{4} \div 1\frac{1}{2} \times 7\frac{1}{2} &= \frac{3}{4} \div \frac{3}{2} \times \frac{15}{2} \\
&= \frac{3}{4} \times \frac{2}{3} \times \frac{15}{2} \\
&= \frac{15}{4} \\
&= 3\frac{3}{4}
\end{aligned}
$$

p.

$$
\begin{aligned}
4\frac{2}{3} \div \frac{3}{2} \times 1\frac{1}{2} \div 2 &= \frac{14}{3} \div \frac{3}{2} \times \frac{3}{2} \div \frac{2}{1} \\
&= \frac{14}{3} \times \frac{2}{3} \times \frac{3}{2} \times \frac{1}{2} \\
&= \frac{7}{3} \\
&= 2\frac{1}{3}
\end{aligned}
$$

5. a.

$$
\begin{aligned}
\frac{1}{2} + \frac{2}{3} \times \frac{3}{8} + \frac{1}{3} &= \frac{1}{2} + \frac{1}{4} + \frac{1}{3} \\
&= \frac{6 + 3 + 4}{12} \\
&= \frac{13}{12} \\
&= 1\frac{1}{12}
\end{aligned}
$$

b.

$$\frac{2}{3} \times \frac{1}{2} + \frac{3}{5} = \frac{1}{3} + \frac{3}{5}$$
$$= \frac{5 + 9}{15}$$
$$= \frac{14}{15}$$

c.

$$\frac{1}{8} + \frac{2}{3} \times \frac{3}{5} \div \frac{3}{5} = \frac{1}{8} + \frac{2}{3} \times \frac{3}{5} \times \frac{5}{3}$$
$$= \frac{1}{8} + \frac{2}{3}$$
$$= \frac{3 + 16}{24}$$
$$= \frac{19}{24}$$

d.

$$\frac{2}{15} \times \frac{3}{2} \times \frac{1}{4} + 1\frac{2}{3} \div 3 \times \frac{1}{2} = \frac{1}{20} + \frac{5}{3} \div \frac{3}{1} \times \frac{1}{2}$$
$$= \frac{1}{20} + \frac{5}{3} \times \frac{1}{3} \times \frac{1}{2}$$
$$= \frac{1}{20} + \frac{5}{18}$$
$$= \frac{9 + 50}{180}$$
$$= \frac{59}{180}$$

e.

$$3 \div \frac{4}{7} - 2\frac{1}{4} + \frac{2}{5} \times 4\frac{1}{2} \div 8 = \frac{3}{1} \div \frac{4}{7} - \frac{9}{4} + \frac{2}{5} \times \frac{9}{2} \div \frac{8}{1}$$
$$= \frac{3}{1} \times \frac{7}{4} - \frac{9}{4} + \frac{2}{5} \times \frac{9}{2} \times \frac{1}{8}$$
$$= \frac{21}{4} - \frac{9}{4} + \frac{9}{40}$$
$$= \frac{210 - 90 + 9}{40}$$
$$= \frac{129}{40}$$
$$= 3\frac{9}{40}$$

f.

$$\frac{3}{4} \div \frac{2}{3} - \frac{1}{2} \times \frac{3}{5} + \frac{2}{3} \times \frac{1}{5} = \frac{3}{4} \times \frac{3}{2} - \frac{3}{10} + \frac{2}{15}$$
$$= \frac{9}{8} - \frac{3}{10} + \frac{2}{15}$$
$$= \frac{135 - 36 + 16}{120}$$
$$= \frac{115}{120}$$
$$= \frac{23}{24}$$

g.

$$\frac{1}{3} \times \frac{2}{5} + 2 - \frac{3}{4} \div \frac{5}{4} = \frac{1}{3} \times \frac{2}{5} + \frac{2}{1} - \frac{3}{4} \times \frac{4}{5}$$
$$= \frac{2}{15} + \frac{2}{1} - \frac{3}{5}$$
$$= \frac{2 + 30 - 9}{15}$$
$$= \frac{23}{15}$$
$$= 1\frac{8}{15}$$

h.

$$\frac{1}{2}\left(2 - \frac{1}{3}\right) + \frac{3}{4}\left(4 + \frac{1}{2}\right) = \frac{1}{2}\left(\frac{2}{1} - \frac{1}{3}\right) + \frac{3}{4}\left(\frac{4}{1} + \frac{1}{2}\right)$$
$$= \frac{1}{2} \times \frac{6 - 1}{3} + \frac{3}{4} \times \frac{8 + 1}{2}$$
$$= \frac{1}{2} \times \frac{5}{3} + \frac{3}{4} \times \frac{9}{2}$$
$$= \frac{5}{6} + \frac{27}{8}$$
$$= \frac{20 + 81}{24}$$
$$= \frac{101}{24}$$
$$= 4\frac{5}{24}$$

i.

$$4\frac{1}{5} \times \frac{5}{7} \div \frac{3}{4} - \frac{2}{3}\left(\frac{1}{5} + \frac{3}{4} \times \frac{1}{2}\right) = \frac{21}{5} \times \frac{5}{7} \div \frac{3}{4} - \frac{2}{3}\left(\frac{1}{5} + \frac{3}{8}\right)$$

$$= \frac{21}{5} \times \frac{5}{7} \times \frac{4}{3} - \frac{2}{3} \times \frac{8 + 15}{40}$$

$$= \frac{4}{1} - \frac{2}{3} \times \frac{23}{40}$$

$$= \frac{4}{1} - \frac{23}{60}$$

$$= \frac{240 - 23}{60}$$

$$= \frac{217}{60}$$

$$= 3\frac{37}{60}$$

j.

$$2\left(\frac{1}{2}\right)\left(\frac{3}{7} \times \frac{1}{2} + \frac{1}{3}\right) = \frac{2}{1}\left(\frac{1}{2}\right)\left(\frac{3}{14} + \frac{1}{3}\right)$$

$$= \frac{2}{1} \times \frac{1}{2} \times \frac{9 + 14}{42}$$

$$= \frac{2}{1} \times \frac{1}{2} \times \frac{23}{42}$$

$$= \frac{23}{42}$$

k.

$$\frac{3}{5} \times \frac{2}{3} - \frac{1}{10}\left(\frac{2}{5} - \frac{1}{2} \times \frac{1}{3}\right) + \frac{3}{4} \times \frac{2}{5} = \frac{2}{5} - \frac{1}{10}\left(\frac{2}{5} - \frac{1}{6}\right) + \frac{3}{10}$$

$$= \frac{2}{5} - \frac{1}{10} \times \frac{12 - 5}{30} + \frac{3}{10}$$

$$= \frac{2}{5} - \frac{1}{10} \times \frac{7}{30} + \frac{3}{10}$$

$$= \frac{2}{5} - \frac{7}{300} + \frac{3}{10}$$

$$= \frac{120 - 7 + 90}{300}$$

$$= \frac{203}{300}$$

l.

$$\frac{1}{3} + \frac{2}{3}\left(\frac{3}{4} + 1\right) + 2\left(\frac{3}{5} \times \frac{2}{3} + 3 \div \frac{12}{7}\right)$$

$$= \frac{1}{3} + \frac{2}{3}\left(\frac{3}{4} + \frac{1}{1}\right) + \frac{2}{1}\left(\frac{2}{5} + \frac{3}{1} \div \frac{12}{7}\right)$$

$$= \frac{1}{3} + \frac{2}{3} \times \frac{3 + 4}{4} + \frac{2}{1}\left(\frac{2}{5} + \frac{3}{1} \times \frac{7}{12}\right)$$

$$= \frac{1}{3} + \frac{2}{3} \times \frac{7}{4} + \frac{2}{1}\left(\frac{2}{5} + \frac{7}{4}\right)$$

$$= \frac{1}{3} + \frac{7}{6} + \frac{2}{1} \times \frac{8 + 35}{20}$$

$$= \frac{1}{3} + \frac{7}{6} + \frac{2}{1} \times \frac{43}{20}$$

$$= \frac{1}{3} + \frac{7}{6} + \frac{43}{10}$$

$$= \frac{10 + 35 + 129}{30}$$

$$= \frac{174}{30}$$

$$= \frac{29}{5}$$

$$= 5\frac{4}{5}$$

m.

$$\left(\frac{1}{4} \times \frac{6}{7} + \frac{1}{2}\right)\left(\frac{5}{6} - 2 \times \frac{1}{8}\right) = \left(\frac{3}{14} + \frac{1}{2}\right)\left(\frac{5}{6} - \frac{2}{1} \times \frac{1}{8}\right)$$

$$= \frac{3 + 7}{14}\left(\frac{5}{6} - \frac{1}{4}\right)$$

$$= \frac{10}{14} \times \frac{10 - 3}{12}$$

$$= \frac{10}{14} \times \frac{7}{12}$$

$$= \frac{5}{12}$$

n.

$$\frac{2}{-3} - \frac{-4}{5} \times \frac{-5}{-6} + \frac{-1}{3} = \frac{-2}{3} + \frac{4}{5} \times \frac{5}{6} - \frac{1}{3}$$

$$= \frac{-2}{3} + \frac{2}{3} - \frac{1}{3}$$

$$= \frac{-2 + 2 - 1}{3}$$

$$= \frac{-1}{3}$$

$$= -\frac{1}{3}$$

o.

$$\frac{-6}{7} \times \frac{-1}{2} + \frac{1}{-4} = \frac{3}{7} - \frac{1}{4}$$
$$= \frac{12 - 7}{28}$$
$$= \frac{5}{28}$$

p.

$$\frac{-2}{-5} - \frac{-2}{3} \times \frac{3}{4} \times \frac{-2}{-3} = \frac{2}{5} + \frac{1}{3}$$
$$= \frac{6 + 5}{15}$$
$$= \frac{11}{15}$$

q.

$$\frac{-5}{6} \times \frac{-3}{2} \times \frac{-1}{-5} - \frac{-3}{5} \div (-6) \times \frac{-5}{2} = \frac{1}{4} + \frac{3}{5} \div \frac{6}{1} \times \frac{5}{2}$$
$$= \frac{1}{4} + \frac{3}{5} \times \frac{1}{6} \times \frac{5}{2}$$
$$= \frac{1}{4} + \frac{1}{4}$$
$$= \frac{1 + 1}{4}$$
$$= \frac{2}{4}$$
$$= \frac{1}{2}$$

r.

$$-\frac{1}{3} \div \frac{5}{-6} - 3\frac{1}{2} + \frac{-1}{-3} \times \frac{-3}{1} \div \frac{-2}{5} = \frac{1}{3} \div \frac{5}{6} - \frac{7}{2} + \frac{1}{3} \times \frac{3}{1} \div \frac{2}{5}$$
$$= \frac{1}{3} \times \frac{6}{5} - \frac{7}{2} + \frac{1}{3} \times \frac{3}{1} \times \frac{5}{2}$$
$$= \frac{2}{5} - \frac{7}{2} + \frac{5}{2}$$
$$= \frac{4 - 35 + 25}{10}$$
$$= \frac{-6}{10}$$
$$= -\frac{3}{5}$$

s.

$$\frac{-5}{-7} \div \frac{1}{-7} - \frac{2}{-5} \times \frac{-5}{3} - \frac{2}{7} \times 1\frac{2}{5} = -\frac{5}{7} \div \frac{1}{7} - \frac{2}{5} \times \frac{5}{3} - \frac{2}{7} \times \frac{7}{5}$$

$$= -\frac{5}{7} \times \frac{7}{1} - \frac{2}{3} - \frac{2}{5}$$

$$= -\frac{5}{1} - \frac{2}{3} - \frac{2}{5}$$

$$= \frac{-5}{1} - \frac{2}{3} - \frac{2}{5}$$

$$= \frac{-75 - 10 - 6}{15}$$

$$= \frac{-91}{15}$$

$$= -6\frac{1}{15}$$

t.

$$\frac{1}{3} \times \frac{-6}{5} - \frac{-2}{-7} + \frac{1}{-4} \div \frac{-5}{4} = -\frac{1}{3} \times \frac{6}{5} - \frac{2}{7} + \frac{1}{4} \div \frac{5}{4}$$

$$= -\frac{2}{5} - \frac{2}{7} + \frac{1}{4} \times \frac{4}{5}$$

$$= \frac{-2}{5} - \frac{2}{7} + \frac{1}{5}$$

$$= \frac{-14 - 10 + 7}{35}$$

$$= \frac{-17}{35}$$

$$= -\frac{17}{35}$$

6. a. Numerator: $8 - \frac{2}{3}$
   Denominator: 4

b. Numerator: $\frac{9}{2} - 2$
   Denominator: $3\frac{1}{4}$

c. Numerator: 2
   Denominator: $\frac{3}{4} \times \frac{1}{3} + \frac{1}{2} \times 5$

d. Numerator: $5 + 3\frac{1}{2}$
   Denominator: $2 - 1\frac{2}{3}$

7. a.

$$\frac{8 - \frac{2}{3}}{4} = \frac{\frac{8}{1} - \frac{2}{3}}{\frac{4}{1}}$$

$$= \frac{\frac{24 - 2}{3}}{\frac{4}{1}}$$

$$= \frac{\frac{22}{3}}{\frac{4}{1}}$$

$$= \frac{22 \times 1}{3 \times 4}$$

$$= \frac{11}{6}$$

$$= 1\frac{5}{6}$$

b.

$$\frac{-8}{2 + \frac{3}{4}} = \frac{\frac{-8}{1}}{\frac{2}{1} + \frac{3}{4}}$$

$$= \frac{\frac{-8}{1}}{\frac{8 + 3}{4}}$$

$$= \frac{\frac{-8}{1}}{\frac{11}{4}}$$

$$= -\frac{8 \times 4}{1 \times 11}$$

$$= -\frac{32}{11}$$

$$= -2\frac{10}{11}$$

c.

$$\frac{5 + 3\frac{1}{2}}{2 - 1\frac{2}{3}} = \frac{\frac{5}{1} + \frac{7}{2}}{\frac{2}{1} - \frac{5}{3}}$$

$$= \frac{\frac{10 + 7}{2}}{\frac{6 - 5}{3}}$$

$$= \frac{\frac{17}{2}}{\frac{1}{3}}$$

$$= \frac{17 \times 3}{2 \times 1}$$

$$= \frac{51}{2}$$

$$= 25\frac{1}{2}$$

d.

$$\frac{1\frac{1}{2} + 3\frac{1}{4}}{2\frac{1}{3} - 1\frac{1}{2}} = \frac{\frac{3}{2} + \frac{13}{4}}{\frac{7}{3} - \frac{3}{2}}$$

$$= \frac{\frac{6 + 13}{4}}{\frac{14 - 9}{6}}$$

$$= \frac{\frac{19}{4}}{\frac{5}{6}}$$

$$= \frac{19 \times 6}{4 \times 5}$$

$$= \frac{57}{10}$$

$$= 5\frac{7}{10}$$

e.

$$\frac{\frac{9}{2} - 2}{3\frac{1}{4}} = \frac{\frac{9}{2} - \frac{2}{1}}{\frac{13}{4}}$$

$$= \frac{\frac{9 - 4}{2}}{\frac{13}{4}}$$

$$= \frac{\frac{5}{2}}{\frac{13}{4}}$$

$$= \frac{5 \times 4}{2 \times 13}$$

$$= \frac{10}{13}$$

f.

$$\frac{-6\frac{2}{3} + 4 \times \frac{3}{2}}{2} = \frac{-\frac{20}{3} + \frac{6}{1}}{\frac{2}{1}}$$

$$= \frac{-\frac{20}{3} + \frac{6}{1}}{\frac{2}{1}}$$

$$= \frac{\frac{-20 + 18}{3}}{\frac{2}{1}}$$

$$= \frac{\frac{-2}{3}}{\frac{2}{1}}$$

$$= -\frac{2 \times 1}{3 \times 2}$$

$$= -\frac{1}{3}$$

g.

$$\frac{5 - 2\frac{1}{2}}{\frac{3}{4} + 2 \times \frac{1}{3}} = \frac{\frac{5}{1} - \frac{5}{2}}{\frac{3}{4} + \frac{2}{3}}$$

$$= \frac{\frac{10 - 5}{2}}{\frac{9 + 8}{12}}$$

$$= \frac{\frac{5}{2}}{\frac{17}{12}}$$

$$= \frac{5 \times 12}{2 \times 17}$$

$$= \frac{30}{17}$$

$$= 1\frac{13}{17}$$

h.

$$\frac{6\frac{1}{3} - 4}{\frac{2}{5} + \frac{1}{3} \times 6} = \frac{\frac{19}{3} - \frac{4}{1}}{\frac{2}{5} + \frac{2}{1}}$$

$$= \frac{\frac{19 - 12}{3}}{\frac{2 + 10}{5}}$$

$$= \frac{\frac{7}{3}}{\frac{12}{5}}$$

$$= \frac{7 \times 5}{3 \times 12}$$

$$= \frac{35}{36}$$

i.

$$\frac{-4\frac{1}{3} - 4 \times \frac{3}{8}}{-4 - 1\frac{1}{2}} = \frac{-\frac{13}{3} - \frac{4}{1} \times \frac{3}{8}}{\frac{-4}{1} - \frac{3}{2}}$$

$$= \frac{-\frac{13}{3} - \frac{3}{2}}{\frac{-8 - 3}{2}}$$

$$= \frac{\frac{-26 - 9}{6}}{\frac{-11}{2}}$$

$$= \frac{\frac{-35}{6}}{\frac{-11}{2}}$$

$$= \frac{35 \times 2}{6 \times 11}$$

$$= \frac{35}{33}$$

$$= 1\frac{2}{33}$$

j.

$$\frac{8 - 1}{\frac{3}{4} \times 5} = \frac{7}{\frac{3}{4} \times \frac{5}{1}}$$

$$= \frac{\frac{7}{1}}{\frac{15}{4}}$$

$$= \frac{7 \times 4}{1 \times 15}$$

$$= \frac{28}{15}$$

$$= 1\frac{13}{15}$$

k.

$$\frac{4 - 1\frac{1}{3}}{2 \times \frac{1}{2} + 1} = \frac{\frac{4}{1} - \frac{4}{3}}{\frac{2}{1} \times \frac{1}{2} + \frac{1}{1}}$$

$$= \frac{\frac{12 - 4}{3}}{\frac{1}{1} + \frac{1}{1}}$$

$$= \frac{\frac{8}{3}}{\frac{1 + 1}{1}}$$

$$= \frac{\frac{8}{3}}{\frac{2}{1}}$$

$$= \frac{8 \times 1}{3 \times 2}$$

$$= \frac{4}{3}$$

$$= 1\frac{1}{3}$$

l.

$$\frac{2}{\frac{3}{4} + \frac{1}{2} \times 5} = \frac{\frac{2}{1}}{\frac{3}{4} + \frac{1}{2} \times \frac{5}{1}}$$

$$= \frac{\frac{2}{1}}{\frac{3}{4} + \frac{5}{2}}$$

$$= \frac{\frac{2}{1}}{\frac{3 + 10}{4}}$$

$$= \frac{\frac{2}{1}}{\frac{13}{4}}$$

$$= \frac{2 \times 4}{1 \times 13}$$

$$= \frac{8}{13}$$

m.

$$\frac{\frac{3}{4} + \frac{1}{2} \times \frac{4}{3}}{2 + \frac{1}{3}} = \frac{\frac{3}{4} + \frac{2}{3}}{\frac{2}{1} + \frac{1}{3}}$$

$$= \frac{\frac{9 + 8}{12}}{\frac{6 + 1}{3}}$$

$$= \frac{\frac{17}{12}}{\frac{7}{3}}$$

$$= \frac{17 \times 3}{12 \times 7}$$

$$= \frac{17}{28}$$

n.

$$\frac{\frac{1}{3} \times 4 + 2}{3 - \frac{1}{2} \times \frac{1}{4}} = \frac{\frac{1}{3} \times \frac{4}{1} + \frac{2}{1}}{\frac{3}{1} - \frac{1}{8}}$$

$$= \frac{\frac{4}{3} + \frac{2}{1}}{\frac{24 - 1}{8}}$$

$$= \frac{\frac{4 + 6}{3}}{\frac{23}{8}}$$

$$= \frac{\frac{10}{3}}{\frac{23}{8}}$$

$$= \frac{10 \times 8}{3 \times 23}$$

$$= \frac{80}{69}$$

$$= 1\frac{11}{69}$$

o.

$$\frac{\frac{1}{2} + \frac{2}{3} \times 6}{\frac{3}{4} \div \frac{1}{2} + 1} = \frac{\frac{1}{2} + \frac{2}{3} \times \frac{6}{1}}{\frac{3}{4} \times \frac{2}{1} + \frac{1}{1}}$$

$$= \frac{\frac{1}{2} + \frac{4}{1}}{\frac{3}{2} + \frac{1}{1}}$$

$$= \frac{\frac{1+8}{2}}{\frac{3+2}{2}}$$

$$= \frac{\frac{9}{2}}{\frac{5}{2}}$$

$$= \frac{9 \times 2}{2 \times 5}$$

$$= \frac{9}{5}$$

$$= 1\frac{4}{5}$$

p.

$$\frac{2\frac{1}{4} - \frac{1}{2} \times \frac{1}{3}}{\frac{3}{4} + \frac{1}{2}} = \frac{\frac{9}{4} - \frac{1}{6}}{\frac{3+2}{4}}$$

$$= \frac{\frac{27-2}{12}}{\frac{5}{4}}$$

$$= \frac{\frac{25}{12}}{\frac{5}{4}}$$

$$= \frac{25 \times 4}{12 \times 5}$$

$$= \frac{5}{3}$$

$$= 1\frac{2}{3}$$

Exercise Set 4.3.6

(a)

(b)

(c)

(d)

(e)

(f)

(g)

1.

(h)

(i)

(j)

(k)

(l)

2. a. $\frac{1}{4}$　　　　c. $\frac{1}{3}$　　　　e. $-\frac{3}{8}$　　　　g. $-\frac{4}{5}$

　　b. $\frac{5}{8}$　　　　d. $\frac{1}{5}$　　　　f. $-\frac{3}{4}$　　　　h. $-\frac{2}{3}$

3. a. $2\frac{3}{5}$　　　　　c. $15\frac{2}{5}$　　　　　e. $-16\frac{5}{8}$

　　b. $3\frac{3}{8}$　　　　　d. $-5\frac{2}{5}$　　　　　f. $-5\frac{1}{6}$

Exercise Set 4.4.1

1. a. 1 to the power of 8　　　　d. 18 cubed
　 b. 0 to the power of 12　　　 e. 9 to the power of 1
　 c. 4 squared

2. a. 4^7 b. 6^2 c. 12^5 d. 2^3

3. a. base: 7   b. base: 1   c. base: 15
    exponent: 2   exponent: 75   exponent: 6

4. a. 2, 4, 8, 16   b. 5, 25, 125, 625

5. a. 8   d. 196   g. 1 000 000   j. 256
   b. 243   e. 24   h. 0   k. 90 000
   c. 1   f. 10 000   i. 15^{28} l. 8000

6. a. 0   e. 2 744 000   i. 1 000 000
   b. 100 000   f. 1   j. 1 440 000
   c. 1   g. 0   k. 1
   d. 10 000   h. 810 000   l. indeterminate

7. a. 81   d. -49 g. -125
   b. -1 e. -1000 h. -27
   c. -125 f. -16 i. 64

8. a. $\dfrac{16}{625}$ d. $5\dfrac{19}{25}$ g. $42\dfrac{7}{8}$ j. $11\dfrac{25}{64}$

   b. $7\dfrac{19}{32}$ e. $\dfrac{1}{10\,000}$ h. $\dfrac{81}{10\,000}$ k. $19\dfrac{683}{1000}$

   c. $28\dfrac{4}{9}$ f. $2\dfrac{7}{9}$ i. $\dfrac{343}{27\,000\,000}$ l. $\dfrac{25}{196}$

9. a. $1\dfrac{1}{7}$ d. $\dfrac{16}{125}$ g. $\dfrac{1}{2}$

   b. $\dfrac{1}{16}$ e. $\dfrac{1}{100}$ h. $8\dfrac{1}{10}$

   c. $10\dfrac{2}{3}$ f. $10\dfrac{1}{8}$ i. $12\dfrac{28}{81}$

10. a. $\dfrac{16}{81}$ d. $-\dfrac{16}{25}$ g. $-\dfrac{343}{27\,000}$

    b. $-\dfrac{11}{20}$ e. $\dfrac{16}{625}$ h. $-\dfrac{1}{10\,000}$

    c. $\dfrac{27}{125}$ f. $-\dfrac{125}{216}$ i. $-\dfrac{1024}{16\,807}$

11. a. 1   c. $\dfrac{1}{5}$ e. $\dfrac{4}{5}$ g. -2
    b. 12   d. 1   f. 1   h. 1

i. 1   j. 1   k. indeterminate   l. 0

12. a. $\frac{1}{16}$ c. $\frac{1}{32}$ e. $-\frac{1}{9}$ g. $-\frac{1}{2}$

    b. $\frac{1}{8}$ d. $\frac{1}{9}$ f. $-\frac{1}{2}$ h. $\frac{1}{64}$

13. a. $3\frac{3}{8}$ c. 4   e. -125 g. 4

    b. $11\frac{1}{9}$ d. $-9\frac{97}{256}$ f. $-\frac{2}{3}$ h. $-3\frac{1}{2}$

Exercise Set 4.4.2

1. a. the cube root of 64   d. the 7th root of 20
    b. the 4th root of 625   e. the square root of 9
    c. the square root of 36   f. the 6th root of 1

2. a. $\sqrt{25}$ c. $\sqrt[4]{81}$ e. $\sqrt{4}$
    b. $\sqrt[5]{1}$ d. $\sqrt[3]{8}$ f. $\sqrt[3]{125}$

3. a. root number: 3   c. root number: 4
       radicand: 8   radicand: 10 000

    b. root number: 2   d. root number: 3
       radicand: 49   radicand: 1

4. a. $64^{\frac{1}{3}}$ c. $24^{\frac{1}{4}}$ e. $9^{\frac{7}{2}}$ g. $400^{\frac{1}{2}}$
    b. $6^{\frac{1}{2}}$ d. $2^{\frac{4}{5}}$ f. $12^{\frac{1}{10}}$ h. $5^{\frac{3}{4}}$

5. a. $\sqrt{5}$ e. $\sqrt[4]{6}$
    b. $\sqrt[3]{12}^2$ or $\sqrt[3]{12^2}$ f. $\sqrt[5]{100}^{12}$ or $\sqrt[5]{100^{12}}$
    c. $\sqrt[5]{1}^3$ or $\sqrt[5]{1^3}$ g. $\sqrt[3]{5}^4$ or $\sqrt[3]{5^4}$
    d. $\sqrt[3]{0}$ h. $\sqrt[5]{7}$

6. a. 4   g. 3   m. undefined   s. -10
    b. 2   h. 10   n. -2 t. $\frac{1}{3}$
    c. 13   i. 10   o. -4 u. $\frac{2}{3}$
    d. 8   j. 2   p. 10   v. $\frac{2}{3}$
    e. 0   k. -2 q. undefined   w. undefined
    f. 2   l. 2   r. 10

x. $\frac{1}{2}$ y. undefined

Exercise Set 4.4.3

1. a. log to the base 6 of 36   d. log to the base 10 of 150
   b. log to the base 3 of 2   e. log to the base 10 of 100
   c. log to the base 2 of 7   f. log to the base 10 of 24

2. a. $\log_3 81$ c. $\log_4 16$ e. $\log_2 1$
   b. $\log 34$ d. $\log_7 7$ f. $\log_5 125$

3. a. 6   b. 10   c. 10   d. 3

4. a. 2   d. 2   g. 1
   b. -4 e. -3 h. -1
   c. 2   f. 3   i. 0

Exercise Set 4.4.4

1. a.

   $3 \times 2 + 4^2 \times 3$

   $6 + 16 \times 3$

   $6 + 48$

   54

   b.

   $5 \times 3^2 + 8 \times 2$

   $5 \times 9 + 16$

   $45 + 16$

   61

   c.

   $-7 \times 4^2 + 3^3 \times 2$

   $-7 \times 16 + 27 \times 2$

   $-112 + 54$

   -58

   d.

   $2^3 \times 5^2 + 8$

   $8 \times 25 + 8$

   $200 + 8$

   208

   e.

   $15 + 2^5 \times 3^3 - 6^2 \div 9$

   $15 + 32 \times 27 - 36 \div 9$

   $15 + 864 - 4$

   875

   f.

   $-4 + (-3)^2 - (-3)(4)$

   $-4 + 9 + 12$

   17

   g.

   $2(-3)(4^2) + 3^2(-2)(-1)$

   $2(-3)(16) + 9(-2)(-1)$

   $-96 + 18$

   -78

   h.

   $2(4^2)(-1) - (2^3)(-3)$

   $2(16)(-1) - 8(-3)$

   $-32 + 24$

   -8

i.

$$3\left(2 + 3^2\right)(1 \times 3 - 2)$$
$$3\left(2 + 9\right)(3 - 2)$$
$$3\left(11\right)(1)$$
$$33$$

j.

$$2\left(4 - 2^3\right) - 3 \times 4^2$$
$$2\left(4 - 8\right) - 3 \times 16$$
$$2\left(-4\right) - 48$$
$$-8 - 48$$
$$-56$$

k.

$$18\left(4 \times 2 - 3^2\right)^2$$
$$18\left(8 - 9\right)^2$$
$$18\left(-1\right)^2$$
$$18 \times 1$$
$$18$$

$$8\left(\frac{1}{4}\right)^2 - \frac{3}{5} \times 4 + (-3)\left(\frac{2}{3}\right)^3$$
$$8 \times \frac{1}{16} - \frac{12}{5} + (-3)\left(\frac{8}{27}\right)$$
$$\frac{1}{2} - \frac{12}{5} - \frac{8}{9}$$
$$\frac{45 - 216 - 80}{90}$$
$$\frac{-251}{90}$$
$$-2\frac{71}{90}$$

l.

$$\left(\frac{2}{3}\right)^2 \times 3 - (-5)\left(\frac{1}{3}\right)^3$$
$$\frac{4}{9} \times 3 - (-5)\left(\frac{1}{9}\right)$$
$$\frac{4}{3} + \frac{5}{9}$$
$$\frac{12 + 5}{9}$$
$$\frac{17}{9}$$
$$1\frac{8}{9}$$

m.

$$\frac{2}{3}\left(\frac{3}{4}\right)^2 + \frac{1}{2} \div \frac{4}{5} - \frac{1}{3}$$
$$\frac{2}{3} \times \frac{9}{16} + \frac{1}{2} \times \frac{5}{4} - \frac{1}{3}$$
$$\frac{3}{8} + \frac{5}{8} - \frac{1}{3}$$
$$\frac{9 + 15 - 8}{24}$$
$$\frac{16}{24}$$
$$\frac{2}{3}$$

n.

$$4 \times 3^2 + 2\,(-1 + 2 \times 3)$$
$$4 \times 9 + 2\,(-1 + 6)$$
$$36 + 2 \times 5$$
$$36 + 10$$
$$46$$

p.

$$\frac{1}{3} \times 2 + \left(\frac{3}{4}\right)^2$$
$$\frac{2}{3} + \frac{9}{16}$$
$$\frac{32 + 27}{48}$$
$$\frac{59}{48}$$
$$1\frac{11}{48}$$

q.

$$- (-3)^2\,(5)\,(2) - 4\,(-2^3)\,(4) - 4^3 \div 8$$
$$- 9\,(5)\,(2) - 4\,(-8)\,(4) - 64 \div 8$$
$$- 90 + 128 - 8$$
$$30$$

r.

$$- 2\,(4 + 2 \times 3) + 2\,(3^4 - 2 \times 5)$$
$$- 2\,(4 + 6) + 2\,(81 - 10)$$
$$- 2\,(10) + 2\,(71)$$
$$- 20 + 142$$
$$122$$

s.

$$(2 + 3 \times 4 - 18)^3 + 4\,(5 - 3 \times 2)^2$$
$$(2 + 12 - 18)^3 + 4\,(5 - 6)^2$$
$$(-4)^3 + 4\,(-1)^2$$
$$- 64 + 4 \times 1$$
$$- 64 + 4$$
$$- 60$$

2. a.

$$4\sqrt{25} + 3\sqrt{4}$$
$$4 \times 5 + 3 \times 2$$
$$20 + 6$$
$$26$$

b.

$$\sqrt{36} + \sqrt{25}$$
$$6 + 5$$
$$11$$

c.

$$3\sqrt{1} + 2\sqrt{16} - 3\sqrt{4}$$
$$3 \times 1 + 2 \times 4 - 3 \times 2$$
$$3 + 8 - 6$$
$$5$$

d.

$$8 + 3\sqrt{4} - 2 \times 5$$
$$8 + 3 \times 2 - 10$$
$$8 + 6 - 10$$
$$4$$

e.

$$4\sqrt{\frac{1}{36}} + \sqrt{\frac{9}{4}} - \frac{3}{5}$$
$$4 \times \frac{1}{6} + \frac{3}{2} - \frac{3}{5}$$
$$\frac{2}{3} + \frac{3}{2} - \frac{3}{5}$$
$$\frac{20 + 45 - 18}{30}$$
$$\frac{47}{30}$$
$$1\frac{17}{30}$$

f.

$$6\sqrt{9} - \sqrt{16} + 5\sqrt{4}$$
$$6 \times 3 - 4 + 5 \times 2$$
$$18 - 4 + 10$$
$$24$$

g.

$$\sqrt{4} \times 5 + 3\sqrt{25} - 3 \times 4$$
$$2 \times 5 + 3 \times 5 - 12$$
$$10 + 15 - 12$$
$$13$$

h.

$$8\sqrt{1} - 2\sqrt{9} - 5\sqrt{16}$$
$$8 \times 1 - 2 \times 3 - 5 \times 4$$
$$8 - 6 - 20$$
$$-18$$

i.

$$5\sqrt{16} + 2\sqrt{0} - 3\sqrt{4} + \sqrt{1}$$
$$5 \times 4 + 2 \times 0 - 3 \times 2 + 1$$
$$20 + 0 - 6 + 1$$
$$15$$

j.

$$2\sqrt{3} - 5\sqrt{3} + \sqrt{3}$$
$$-2\sqrt{3}$$

k.

$$4\sqrt{5} - 2\sqrt{5} - 7\sqrt{5}$$
$$-5\sqrt{5}$$

l.

$$-\sqrt{2} - 4\sqrt{2} + 5\sqrt{2}$$
$$0$$

m.

$$-3\sqrt[3]{2} + 4\sqrt[3]{2} - \sqrt[3]{2}$$
$$0$$

n.

$$\sqrt[4]{10} + 2\sqrt[4]{10} - 6\sqrt[4]{10}$$
$$-3\sqrt[4]{10}$$

o.

$$-\sqrt{5} + 2\sqrt{5} - 3\sqrt{5}$$
$$-2\sqrt{5}$$

3. a.

$\log_2 8 + \log_2 16$

$3 + 4$

7

b.

$\log_3 9 + \log_2 8$

$2 + 3$

5

c.

$\log_5 25 + \log_3 81$

$2 + 4$

6

d.

$-4\log_2 4 + \log_3 9$

$-4 \times 2 + 2$

$-8 + 2$

-6

e.

$-\log 100 + \log_6 36$

$-2 + 2$

0

f.

$\log_3 81 - 2\log_2 2$

$4 - 2 \times 1$

$4 - 2$

2

g.

$\log_5 \dfrac{1}{25} - (-3)\log_2 \dfrac{1}{2}$

$-2 - (-3)(-1)$

$-2 - 3$

-5

h.

$-\log \dfrac{1}{100} + 2\log_3 3$

$-(-2) + 2 \times 1$

$2 + 2$

4

i.

$-3\log 100 - \log_2 2 - 2\log_3 27$

$-3 \times 2 - 1 - 2 \times 3$

$-6 - 1 - 6$

-13

j.

$4\log_3 3 + 5\log_2 32 - \log 10$

$4 \times 1 + 5 \times 5 - 1$

$4 + 25 - 1$

28

4. a.

$4^2 - 3\log_2 8 + 5\sqrt{16}$

$16 - 3 \times 3 + 5 \times 4$

$16 - 9 + 20$

27

b.

$3\log_4 16 - \sqrt[3]{8}$

$3 \times 2 - 2$

$6 - 2$

4

c.

$4^2 - 2\log 100 + 3\sqrt{25}$

$16 - 2 \times 2 + 3 \times 5$

$16 - 4 + 15$

27

d.

$-2\log_5 5 + \log_3 9 - 2^3$

$-2 \times 1 + 2 - 8$

$-2 + 2 - 8$

-8

e.

$$- \sqrt[3]{81} - 2^4 + \sqrt{36}$$
$$- 4 - 16 + 6$$
$$- 14$$

f.

$$4^2 + 3\log_2 8 - 3\sqrt[3]{-1}$$
$$16 + 3(3) - 3(-1)$$
$$16 + 9 + 3$$
$$28$$

g.

$$\sqrt[3]{-8} - \log_2 \frac{1}{4} + 1$$
$$- 2 - (-2) + 1$$
$$- 2 + 2 + 1$$
$$1$$

h.

$$- 4 + 2\log \frac{1}{100} + 3^2$$
$$- 4 + 2(-2) + 9$$
$$- 4 - 4 + 9$$
$$1$$

i.

$$- 2^3 - \log_2 2 + \log_4 1$$
$$- 8 - 1 + 0$$
$$- 9$$

j. This expression is undefined.

k.

$$\sqrt{2} + 3\sqrt{2}$$
$$4\sqrt{2}$$

l.

$$- 2\sqrt[3]{5} - 4\sqrt[3]{5}$$
$$- 6\sqrt[3]{5}$$

m.

$$4\log_2 3 + \log_2 3 - 3\log_2 3$$
$$2\log_2 3$$

n.

$$3\sqrt{2} - 2\log_2 5 + 5\sqrt{2} - 3\log_2 5$$
$$8\sqrt{2} - 5\log_2 5$$

Exercise Set 5

1. Base-ten notations. The reason is that applied discussions often involve measured values which are rarely exact and lack significance. As a result, the shortcomings of base-ten notations are not of concern while their ease of use is deemed attractive.

2. The use of base-ten notations implies that the values being represented are measured and are, therefore, approximate. Since science formulas demand exactness, the use of measured values in their expressions would make the formulas approximate.

3. The appearance of a numerical value in base-ten form in a science formula implies that the discussion is applied.

4. The use of formal notation in the communication of science formulas indicates that the discussion is theoretical. Appearance of base-ten notations signals the shift from theory to applications of theory.

Exercise Set 6.1.1

1. a. whole part: 926
      fractional part: 4 tenths or $\frac{4}{10}$

   b. whole part: 3
      fractional part: 68 thousandths or $\frac{68}{1000}$

   c. whole part: 95
      fractional part: 0 (There are no bits and pieces.)

   d. whole part: 0 (There are no wholes)
      fractional part: 159 thousandths or $\frac{159}{1000}$

   e. whole part: 0 (There are no wholes)
      fractional part: 4 thousandths or $\frac{4}{1000}$

   f. whole part: 420
      fractional part: 0 (There are no bits and pieces.)

2. The fractional part of a decimal number is always less than 1 in size.

3. a. 1   c. 2   e. 2   g. 0   i. 0
   b. 3   d. 1   f. 2   h. 3

4. 10   9. 10   14. 10 000 times
5. 100   10. 100   15. 10 times
6. 1000   11. 10 times   16. 1000 times
7. 10   12. 100 times
8. 100   13. 100 times

17. a. $4\frac{9}{50}$ c. $\frac{1}{20}$ e. $75\frac{1}{5}$ g. $\frac{3}{10}$ i. $\frac{3}{10}$

    b. $60\frac{7}{100}$ d. $\frac{9}{20}$ f. $\frac{3}{10}$ h. $\frac{3}{10}$

18. a. tenths   d. hundredths   g. thousandths
    b. tens   e. hundreds   h. ten thousandths
    c. ones or units   f. millionths

19. hundredths   23. hundred thousandths
20. thousandths   24. hundredths
21. hundredths   25. ten thousandths
22. tenths

26. a. $0 \times 1 + 4 \times \frac{1}{10} + 5 \times \frac{1}{100} + 7 \times \frac{1}{1000}$

       0 units, 4 tenths, 5 hundredths and 7 thousandths

    b. $0 \times 1 + 0 \times \frac{1}{10} + 7 \times \frac{1}{100}$

       0 units, 0 tenths and 7 hundredths

c. $0 \times 1 + 1 \times \frac{1}{10} + 2 \times \frac{1}{100} + 8 \times \frac{1}{1000} + 4 \times \frac{1}{10000}$

0 units, 1 tenth, 2 hundredths, 8 thousandths and 4 ten thousansdths

d. $9 \times 10 + 2 \times 1 + 7 \times \frac{1}{10} + 9 \times \frac{1}{100}$

9 tens, 2 units, 7 tenths and 9 hundredths

e. $0 \times 1 + 3 \times \frac{1}{10} + 2 \times \frac{1}{100} + 1 \times \frac{1}{1000} + 8 \times \frac{1}{10000} + 5 \times \frac{1}{100\,000}$

0 units, 3 tenths, 2 hundredths, 1 thousandth, 8 ten thousandths and 5 hundred thousansdths

f. $1 \times 100 + 2 \times 10 + 8 \times 1 + 9 \times \frac{1}{10} + 1 \times \frac{1}{100}$

1 hundred, 2 tens, 8 units, 9 tenths and 1 hundredth

Exercise Set 6.1.2

1. a. one hundred twenty and twenty-seven hundredths
   b. negative one and seven tenths
   c. fourteen and six hundredths
   d. forty-five and seven hundred nineteen thousandths
   e. thirty-four thousandths
   f. negative seven hundred fifty-one thousandths
   g. one tenth
   h. three hundredths
   i. negative nine hundred twenty and five thousandths
   j. negative eighteen hundredths
   k. seventy-two and four hundred twenty-nine thousandths
   l. seventy-five ten thousandths

2. a. -56.5 c. 925.076 e. 0.840 g. -4.01
   b. 7.24 d. -0.25 f. 915.759 h. 0.9004

Exercise Set 6.1.3

1. $40.25, 41.7, 42.75, 45.994$
2. $0.68, 0.608, 0.068, 0.0068$
3. $570.254, 570.2504, 570.245, 570.2405, 570.2054$
4. $0.1096, 0.168, 0.1993, 0.21$
5. $-4.1, -3.8, -2.4, -1.5$
6. $15.4, 14.2, -10.1, -13.8$
7. $0, -0.045, -0.08, -0.2, -0.43$
8. $-2.6, -1.34, -1.3, 1.2, 1.25$

Exercise Set 6.1.4

1. a. 39.8   e. 0.69   i. 61   m. 11.4
   b. 6.7   f. 0.588   j. 6
   c. 9.39   g. 0.56   k. 1154.8
   d. 496.29   h. 0.008   l. 118.31

2. a. 0.471   f. 19.274   k. 0.5863   p. 618.54
   b. 0.964   g. 182.74   l. 0.528   q. 94.65
   c. 0.596   h. 96.68   m. 43.05
   d. 0.789   i. 0.926   n. 10.19
   e. 37.77   j. 0.287   o. 16.308

3. a. 0.7   f. 45.5   k. 0.0016   p. 2817.879
   b. 0.06   g. 635.4   l. 205.6   q. 0.0088
   c. 0.041   h. 29.54   m. 9.51   r. 14.68
   d. 0.1   i. 6.447   n. 70.54
   e. 2.45   j. 0.172   o. 22.8

4. a. 9.07   d. 1.092   g. 38.7
   b. 3.33   e. 10.161   h. 242.17
   c. 0.537   f. 252.979   i. 20.92

5. a. 7.68   f. 400   k. 3.82   p. 15.06
   b. 4.08   g. 40   l. 0.382   q. 19.2
   c. 47.73   h. 4   m. 0.0382   r. 4.9
   d. 0.6237   i. 0.4   n. 0.003 82   s. 55.65
   e. 36.205   j. 38.2   o. 131.6   t. 142.56

6. a. 15.6   h. 603.2   o. $0.\overline{27}$ v. $237.\overline{7}$
   b. 2.56   i. 9.84   p. $0.0\overline{23}$ w. 16 000
   c. 78.9   j. 0.25   q. $0.\overline{142857}$ x. 20
   d. 124   k. 0.4   r. 0.375   y. 59.4
   e. 13 200   l. 15   s. $0.1\overline{3}$ z. $0.21\overline{6}$
   f. 4000   m. $0.\overline{6}$ t. 130
   g. 16.2   n. $0.\overline{2}$ u. 0.53

7. a. 1363.2   d. 32.309 523 . . .   g. 0.534 166 . . .
   b. 60.672   e. 1914   h. 369.648
   c. 3.478 222 . . .   f. 2.56   i. 13 800

Exercise Set 6.1.5

1. a.

$$9.4 \times 3.2 + 2.8 \times 4 = 30.08 + 11.2$$
$$= 41.28$$

b.

$$3.6 + 2.7 \times 9.1 = 3.6 + 24.57$$
$$= 28.17$$

c.

$$42.8 - 3.5 \times 2 + 6 \div 1.5 = 42.8 - 7 + 4$$
$$= 39.8$$

d.

$$-18 - 2.46 \times 42 \div 1.5 + (-2.4) \div 0.3 = -18 - 68.88 - 8$$
$$= -94.88$$

e.

$$42.5 \div 0.5 \times 4 + 6.2 \times 4.5 \div 2.5 = 340 + 11.16$$
$$= 351.16$$

f.

$$\frac{18.6 \times 1.4}{2.1} + 8.2 \times 5.6 = 12.4 + 45.92$$
$$= 58.32$$

g.

$$\frac{-19.4\,(4.6)}{-2.3\,(-0.5)} + \frac{-2.3\,(8)}{4\,(0.2)} = -77.6 - 23$$
$$= -100.6$$

h.

$$-4.5\,(-3.6 + 2.8 \times 1.4) = -4.5\,(-3.6 + 3.92)$$
$$= -4.5 \times 0.32$$
$$= -1.44$$

i.

$$6.2\,(5.6 \times 2 + 1.5) - 2.1\,(0.6 - 5 \times 0.1)$$
$$= 6.2\,(11.2 + 1.5) - 2.1\,(0.6 - 0.5)$$
$$= 6.2 \times 12.7 - 2.1 \times 0.1$$
$$= 78.74 - 0.21$$
$$= 78.53$$

j.

$$\frac{18.6 \times 2.5}{5} + \frac{3.2}{4}\left(9.5 + \frac{3}{0.5}\right) = 9.3 + \frac{3.2}{4}\,(9.5 + 6)$$
$$= 9.3 + \frac{3.2}{4} \times 15.5$$
$$= 9.3 + 12.4$$
$$= 21.7$$

k.

$$\frac{16.2}{-2.5} - 3\left(4 \times \frac{2}{0.5} - 3 \times 2.6\right) = -6.48 - 3(16 - 7.8)$$
$$= -6.48 - 3 \times 8.2$$
$$= -6.48 - 24.6$$
$$= -31.08$$

l.

$$3.4\sqrt{1.8} - 2.1\sqrt{5.7} = 3.4 \times 1.341\,640\ldots - 2.1 \times 2.387\,467\ldots$$
$$= 4.561\,578\ldots - 5.013\,681\ldots$$
$$= -0.452\,102\ldots$$

m.

$$-1.7\sqrt{9.93} + 1.17\sqrt{1.1} = -1.7 \times 3.151\,190\ldots + 1.17 \times 1.048\,808\ldots$$
$$= -5.357\,023\ldots + 1.227\,106\ldots$$
$$= -4.129\,917\ldots$$

n.

$$-2.5\sqrt{53.2} - (-4.6)\sqrt{3.1} = -2.5\,(7.293\,833\ldots) - (-4.6)\,(1.760\,681\ldots)$$
$$= -18.234\,582\ldots + 8.099\,135\ldots$$
$$= -10.135\,446\ldots$$

o.

$$4.4 \times 1.7^2 + 2.7 \times 3.8^2 = 4.4 \times 2.89 + 2.7 \times 14.44$$
$$= 12.716 + 38.988$$
$$= 51.704$$

p.

$$-7.4 \times 4.6^3 + 3.2^2 \times 1.6^2 = -7.4 \times 97.336 + 10.24 \times 2.56$$
$$= -720.2864 + 26.2144$$
$$= -694.072$$

q.

$$1.2 \times 3.4^2 \times \sqrt{3.2} = 1.2 \times 11.56 \times 1.788\,854\ldots$$
$$= 24.814\,987\ldots$$

r.

$$\frac{2.5}{3.1} \times \sqrt{4.6} + \frac{1.5}{3.2^2} \times 1.8^3 = \frac{2.5}{3.1} \times 2.144\,761\ldots + \frac{1.5}{10.24} \times 5.832$$
$$= 1.729\,646\ldots + 0.854\,296\ldots$$
$$= 2.583\,942\ldots$$

s.

$$\frac{1.4^3 + 3.2^2}{3.8 \times 4.1^2} = \frac{2.744 + 10.24}{3.8 \times 16.81}$$
$$= \frac{12.984}{3.8 \times 16.81}$$
$$= 0.203\,262\ldots$$

t.

$$3\log_{2.3} 8.4 \;-\; 2\log_{4.5} 17.8 \;=\; 3 \times 2.555\,178\ldots - 2 \times 1.914\,262\ldots$$
$$= 7.665\,536\ldots - 3.828\,524\ldots$$
$$= 3.837\,011\ldots$$

u.

$$-4\log_{6.3} 12.7 \;+\; (-3)\log_{1.4} 0.8$$
$$= -4\,(1.380\,892\ldots) + (-3)\,(-0.663\,185\ldots)$$
$$= -5.523\,571\ldots + 1.989\,556\ldots$$
$$= -3.534\,014\ldots$$

v.

$$-\log 132.7 \;+\; 3\sqrt{5.9} \;=\; -2.122\,870\ldots + 3 \times 2.428\,991\ldots$$
$$= -2.122\,870\ldots + 7.286\,974\ldots$$
$$= 5.164\,103\ldots$$

Exercise Set 6.1.6

1. a. 50
   b. 30
   c. 90
   d. 100
   e. 320
   f. 430
   g. 320
   h. 810
   i. 2490
   j. 3300
   k. 10
   l. 0

2. a. 100
   b. 400
   c. 3600
   d. 3000
   e. 4100
   f. 3300
   g. 4500
   h. 7400
   i. 12 100
   j. 300
   k. 0
   l. 100

3. a. 3000
   b. 9000
   c. 9000
   d. 10 000
   e. 0
   f. 44 000
   g. 0
   h. 3000
   i. 15 000
   j. 4000
   k. 471 000
   l. 100 000

4. a. 30 000
   b. 920 000
   c. 810 000
   d. 3 200 000
   e. 320 000
   f. 740 000
   g. 10 000
   h. 0
   i. 40 000
   j. 10 000
   k. 82 130 000
   l. 20 000

5. a. 320
    b. 43 000
    c. 1920
    d. 1900
    e. 2000
    f. 0
    g. 71 800
    h. 300 000
    i. 1 000 000
    j. 5000
    k. 130 000 000
    l. 0

6. a. −50
    b. 30
    c. −90
    d. −100
    e. 320
    f. 430
    g. −320
    h. 810
    i. 2490
    j. −3300
    k. 10
    l. 0

7. a. 100
    b. −400
    c. 3600
    d. −3000
    e. −4100
    f. 3300
    g. −4500
    h. 7400
    i. −12 100
    j. 300
    k. 0
    l. −100

8. a. −3000
    b. −9000
    c. −9000
    d. 10 000
    e. 0
    f. 44 000
    g. 0
    h. 3000
    i. −15 000
    j. −4000
    k. −471 000
    l. −100 000

9. a. −30 000
    b. −920 000
    c. 810 000
    d. −3 200 000
    e. −320 000
    f. −740 000
    g. 10 000
    h. 0
    i. 40 000
    j. −10 000
    k. −82 130 000
    l. −20 000

10. a. −320
    b. 43 000
    c. −1920
    d. 1920
    e. −1900
    f. 1900
    g. −2000
    h. 2000
    i. 0
    j. 0
    k. 71 800
    l. −300 000
    m. −1 000 000
    n. −5000
    o. −130 000 000
    p. 0

11. a. 36
    b. 35
    c. 36
    d. 35
    e. 473
    f. 2
    g. 0
    h. 10
    i. 2
    j. 7
    k. 3
    l. 200

12. a. 0.4
    b. 0.4
    c. 0.3
    d. 0.3
    e. 0.8
    f. 1.0
    g. 2.9
    h. 47.6
    i. 32.4
    j. 1.5
    k. 10.0
    l. 62.0

13. a. 3.14
    b. 270.89
    c. 0.08
    d. 0.05
    e. 0.10
    f. 20.00
    g. 0.04
    h. 0.20
    i. 0.03
    j. 0.32
    k. 18.20
    l. 0.00

14. a. 0.245
    b. 0.246
    c. 0.245
    d. 0.245
    e. 0.245
    f. 0.000
    g. 0.320
    h. 42.820
    i. 5.462
    j. 270.000
    k. 1.826
    l. 2.435

15. a. 0.0044
    b. 0.0044
    c. 0.0043
    d. 0.0043
    e. 0.0043
    f. 0.3300
    g. 3.1416
    h. 0.0400
    i. 0.0080
    j. 0.0000
    k. 0.0000
    l. 0.0001

16. a. 30   c. 140   e. 100   g. 50
    b. 30   d. 0   f. 100   h. 50

17. a. 300   b. 1300   c. 0   d. 1300

18. a. 6000   b. 3000   c. 55 000   d. 3000

19. a. 45.7   e. 242   i. 1920.0000   m. 0.0064
    b. 0   f. 1920.00   j. 100   n. 0.028
    c. 42610.000   g. 20   k. 0.64
    d. 1920.0   h. 1920.000   l. 72.820

Exercise Set 6.1.7

1. a. 1.5   c. 3.8   e. −27.6
   b. −0.3   d. 49.1   f. −6.5

2. a. 0.22   d. 0.008   g. 42.62
   b. −0.21   e. 0.000   h. 7.75
   c. −0.035   f. −0.0025   i. 2.416

3. a.

b.

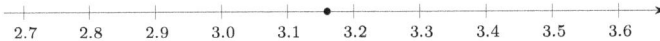

c.

d.

e.

f.

g.

h.

i.

4. Round to the nearest tenth and then graph.
5. Round to the nearest unit and then graph.
6. Round to the nearest hundredth and then graph.
7. Round to the nearest thousandth and then graph.

Exercise Set 6.2.1

1. a. The number is written in scientific notation.

   b. The number is not written in scientific notation. The significand is out of range.

   c. The number is written in scientific notation.

   d. The number is not written in scientific notation. The exponent is not an integer.

   e. The number is not written in scientific notation. The significand is out of range.

   f. The number is written in scientific notation.

2. a. A positive exponent of 10 in scientific notation implies repeated multiplication of the significand by 10.

   b. A 0 exponent of 10 in scientific notation implies multiplication of the significand by 1.

   c. A negative exponent of 10 in scientific notation implies repeated division of the significand by 10.

3. a. 4   d. 8   g. -4

   b. 10   e. 5   h. -8

   c. 2   f. -6 i. 2

4. a. 395 700   d. 0.0001   g. 0.000 019

   b. 621 960   e. 0.03198   h. 90 000

   c. 5200   f. 9.87   i. 10

Exercise Set 6.2.2

1. a. four times ten to the power of three
   b. three times ten to the power of zero
   c. five times ten to the power of negative two
   d. two point four times ten to the power of sixty-two
   e. three point eight one times ten to the power of zero
   f. two point zero zero times ten to the power of negative one
   g. negative three times ten to the power of twenty-one
   h. negative nine times ten to the power of zero
   i. negative five time ten to the power of negative eleven
   j. negative nine point three seven times ten to the power of eight
   k. negative seven point zero times ten to the power of zero
   l. negative one point one one times ten to the power of negative two

2. a. 5×10^8 d. 2.8×10^{12} g. -3×10^{15} j. -8.00×10^{50}

   b. 3×10^0 e. 4.21×10^0 h. -4×10^0 k. -7.16×10^0

   c. 2×10^4 f. 3.0×10^{-45} i. -5×10^{-27} l. -6.921×10^{-7}

Exercise Set 6.2.3

1. a. 4.5×10^3 e. 4.21×10^{-3} i. -2.2×10^3 m. -1.4×10^0

   b. 4.1×10^{-8} f. 3.21×10^1 j. -1.1×10^{-2} n. $-2.450 \quad \times$

   c. -1.7×10^0 g. 7.2×10^7 k. -4.2×10^{-7} 10^{-22}

   d. 3.8×10^5 h. 9.7×10^{-8} l. -2.2×10^6 o. -1.2×10^{42}

2. a. $5.7 \times 10^4, 1.8 \times 10^3, 3.7 \times 10^2$

   b. $2.4 \times 10^8, 5.8 \times 10^5, 3.3 \times 10^0$

   c. $4.7 \times 10^1, 1.8 \times 10^0, 7.2 \times 10^{-1}, 3.4 \times 10^{-2}$

   d. $8 \times 10^{-119}, 4 \times 10^{-120}, 2 \times 10^{-126}$

   e. $4.813 \times 10^8, 4.2 \times 10^8, 3 \times 10^8, 1.99 \times 10^8$

   f. $-1.1 \times 10^2, -3.5 \times 10^3, -2.8 \times 10^4$

   g. $-5.2 \times 10^{-5}, -8 \times 10^{-4}, -8 \times 10^{-2}$

3. a. $-4.2 \times 10^{-8}, -3.81 \times 10^{-8}, -1.9 \times 10^{-8}$

b. -4×10^{14}, -3×10^{14}, -1×10^{14}

c. -8.600×10^{0}, -4.8×10^{0}, -2.71×10^{0}

d. -4.5×10^{-3}, -2.44×10^{-3}, -1.8×10^{-3}, -1×10^{-3}

e. -3.3×10^{3}, -2.7×10^{2}, 4.7×10^{-1}, 1.9×10^{1}

f. -2.8×10^{4}, -1.4×10^{4}, -5×10^{3}, 3×10^{5}

g. -3.8×10^{2}, 1.9×10^{4}, 1.99×10^{4}, 5.9×10^{4}

4. a. The difference in the order of magnitudes of the two numbers is

$$9 - 4 = 5$$

This means that the larger number is anywhere between 10 000 and 1 000 000 times larger than the smaller number.

b. The difference in the order of magnitudes of the two numbers is

$$5 - 3 = 2$$

This means that the larger number is anywhere between 10 to 1000 times larger than the smaller number.

c. The difference in the order of magnitudes of the two numbers is

$$1 - 0 = 1$$

This means that the larger number is anywhere between 1 to 100 times larger than the smaller number.

d. The difference in the order of magnitudes of the two numbers is

$$-3 - (-8) = -3 + 8$$
$$= 5$$

This means that the larger number is anywhere between 10 000 to 1 000 000 times larger than the smaller number.

e. The difference in the order of magnitudes of the two numbers is

$$0 - (-5) = 0 + 5$$
$$= 5$$

This means that the larger number is anywhere between 10 000 to 1 000 000 times larger than the smaller number.

f. The difference in the order of magnitudes of the two numbers is

$$0 - (-3) = 0 + 3$$
$$= 3$$

This means that the larger number is anywhere between 100 to 10 000 times larger than the smaller number.

g. The difference in the order of magnitudes of the two numbers is

$$4 - (-4) = 4 + 4$$
$$= 8$$

This means that the larger number is anywhere between 10^7 to 10^9 times larger than the smaller number.

h. The difference in the order of magnitudes of the two numbers is

$$95 - (-82) = 95 + 82$$
$$= 177$$

This means that the larger number is anywhere between 10^{176} to 10^{178} times larger than the smaller number.

Exercise Set 6.2.4

1. a.

$$4.5 \times 10^3 + 2.1 \times 10^3 = 6.6 \times 10^3$$

b.

$$6.2 \times 10^5 - 7.5 \times 10^5 = -1.3 \times 10^5$$

c.

$$-1.27 \times 10^{23} - 4.7 \times 10^{23} = -5.97 \times 10^{23}$$

d.

$$-1 \times 10^{-4} + 2.7 \times 10^{-4} = 1.7 \times 10^{-4}$$

e.

$$8.235 \times 10^{-4} + 1.8 \times 10^{-2} - 3.9 \times 10^{-3}$$
$$= 0.082\,35 \times 10^{-2} + 1.8 \times 10^{-2} - 0.39 \times 10^{-2}$$
$$= 1.492\,35 \times 10^{-2}$$

f.

$$-2.54 \times 10^3 + 1.6 \times 10^0 - 4.4 \times 10^2$$
$$= -2.54 \times 10^3 + 0.0016 \times 10^3 - 0.44 \times 10^3$$
$$= -2.9784 \times 10^3$$

g.

$$-5 \times 10^{35} - 2.1 \times 10^{32} - 6 \times 10^{33}$$
$$= -5 \times 10^{35} - 0.0021 \times 10^{35} - 0.06 \times 10^{35}$$
$$= -5.0621 \times 10^{35}$$

h.

$$-2.8 \times 10^1 - 3.5 \times 10^2 + 4 \times 10^3$$
$$= -0.028 \times 10^3 - 0.35 \times 10^3 + 4 \times 10^3$$
$$= 3.622 \times 10^3$$

i.

$$4.2 \times 10^{-7} - 3 \times 10^{-5} - 9.92 \times 10^{-5}$$
$$= 0.042 \times 10^{-5} - 3 \times 10^{-5} - 9.92 \times 10^{-5}$$
$$= -12.878 \times 10^{-5}$$
$$= -1.2878 \times 10^{-4}$$

j.

$$-9.5 \times 10^0 + 2.4 \times 10^1 - 1.1 \times 10^{-1}$$
$$= -0.95 \times 10^1 + 2.4 \times 10^1 - 0.011 \times 10^1$$
$$= 1.439 \times 10^1$$

k.

$$3.21 \times 10^4 - 1.9 \times 10^2 + 3 \times 10^2$$
$$= 3.21 \times 10^4 - 0.019 \times 10^4 + 0.03 \times 10^4$$
$$= 3.221 \times 10^4$$

l.

$$-1 \times 10^1 + 1 \times 10^0 - 1 \times 10^{-1}$$
$$= -1 \times 10^1 + 0.1 \times 10^1 - 0.01 \times 10^1$$
$$= -0.91 \times 10^1$$
$$= -9.1 \times 10^0$$

m.

$$3.7 \times 10^{410} + 1.5 \times 10^{408} - 1.2 \times 10^{407}$$
$$= 3.7 \times 10^{410} + 0.015 \times 10^{410} - 0.0012 \times 10^{410}$$
$$= 3.7138 \times 10^{410}$$

n.

$$-6.3 \times 10^{520} - 3.1 \times 10^{519} = -6.3 \times 10^{520} - 0.31 \times 10^{520}$$
$$= -6.61 \times 10^{520}$$

o.

$$2.7 \times 10^{-208} + 3.5 \times 10^{-206} + 2.2 \times 10^{-207}$$
$$= 0.027 \times 10^{-206} + 3.5 \times 10^{-206} + 0.22 \times 10^{-206}$$
$$= 3.747 \times 10^{-206}$$

2. a.

$$\left(7.9 \times 10^4\right)\left(2.1 \times 10^3\right) = 16.59 \times 10^{4+3}$$
$$= 16.59 \times 10^7$$
$$= 1.659 \times 10^8$$

b.

$$\left(1.4 \times 10^{-3}\right)\left(4.9 \times 10^2\right) = 6.86 \times 10^{-3+2}$$
$$= 6.86 \times 10^{-1}$$

c.

$$\left(-7.3 \times 10^{-9}\right)\left(4 \times 10^{-2}\right) = -29.2 \times 10^{-9-2}$$
$$= -29.2 \times 10^{-11}$$
$$= -2.92 \times 10^{-10}$$

d.

$$\left(1.5 \times 10^{-6}\right)\left(9 \times 10^{-5}\right) = 13.5 \times 10^{-6-5}$$
$$= 13.5 \times 10^{-11}$$
$$= 1.35 \times 10^{-10}$$

e.

$$\left(-4.4 \times 10^0\right)\left(-7.11 \times 10^4\right) = 31.284 \times 10^{0+4}$$
$$= 31.284 \times 10^4$$
$$= 3.1284 \times 10^5$$

f.

$$\left(8.4 \times 10^4\right)\left(6.2 \times 10^7\right) = 52.08 \times 10^{4+7}$$
$$= 52.08 \times 10^{11}$$
$$= 5.208 \times 10^{12}$$

g.

$$\left(-3.66 \times 10^{-2}\right)\left(-5.4 \times 10^{-8}\right) = 19.764 \times 10^{-2-8}$$
$$= 19.764 \times 10^{-10}$$
$$= 1.9764 \times 10^{-9}$$

h.

$$\frac{3.72 \times 10^7}{2.7 \times 10^2} = 1.377\ 777\ \dots \times 10^{7-2}$$
$$= 1.377\ 777\ \dots \times 10^5$$

i.

$$\frac{1.8 \times 10^3}{9.4 \times 10^2} = 0.191\ 489 \ldots \times 10^{3-2}$$
$$= 0.191\ 489 \ldots \times 10^1$$
$$= 1.914\ 893 \ldots \times 10^0$$

j.

$$\frac{9.21 \times 10^{-1}}{3.7 \times 10^4} = 2.489\ 189 \ldots \times 10^{-1-4}$$
$$= 2.489\ 189 \ldots \times 10^{-5}$$

k.

$$\frac{3.214 \times 10^{-5}}{2.7 \times 10^{-2}} = 1.190\ 370 \ldots \times 10^{-5+2}$$
$$= 1.190\ 370 \ldots \times 10^{-3}$$

l.

$$\frac{1.8 \times 10^7}{2.59 \times 10^{-5}} = 0.694\ 980 \ldots \times 10^{7+5}$$
$$= 0.694\ 980 \ldots \times 10^{12}$$
$$= 6.949\ 806 \ldots \times 10^{11}$$

m.

$$\left(-9.12 \times 10^4\right)\left(-3.87 \times 10^5\right)\left(1.8 \times 10^{-1}\right)$$
$$= 63.529\ 92 \times 10^{4+5-1}$$
$$= 63.529\ 92 \times 10^8$$
$$= 6.352\ 992 \times 10^9$$

n.

$$\left(-1.2 \times 10^{-5}\right)\left(3.15 \times 10^{-2}\right)\left(1.02 \times 10^{-6}\right)$$
$$= -3.8556 \times 10^{-5-2-6}$$
$$= -3.8556 \times 10^{-13}$$

o.

$$\left(5 \times 10^{-3}\right)\left(1.8 \times 10^2\right)\left(4.1 \times 10^{-2}\right)$$
$$= 36.9 \times 10^{-3+2-2}$$
$$= 36.9 \times 10^{-3}$$
$$= 3.69 \times 10^{-2}$$

p.

$$\frac{\left(4.1 \times 10^{-3}\right)\left(5.7 \times 10^{0}\right)}{\left(7.19 \times 10^{5}\right)\left(2.16 \times 10^{-2}\right)\left(1.1 \times 10^{-1}\right)} = 1.367\ 991\ \dots \times 10^{-3+0-5+2+1}$$

$$= 1.367\ 991\ \dots \times 10^{-5}$$

q.

$$\frac{\left(-9.87 \times 10^{6}\right)\left(-3.8 \times 10^{-6}\right)\left(-4.15 \times 10^{-3}\right)}{\left(1.9 \times 10^{-5}\right)\left(-4.61 \times 10^{-9}\right)\left(-5.6 \times 10^{7}\right)} = -3.173\ 264\ \dots \times 10^{6-6-3+5+9-7}$$

$$= -3.173\ 264\ \dots \times 10^{4}$$

r.

$$\frac{\left(4.1 \times 10^{-2}\right)\left(3.2 \times 10^{-6}\right)\left(4.3 \times 10^{-1}\right)}{9.9 \times 10^{-2}} = 5.698\ 585\ \dots \times 10^{-2-6-1+2}$$

$$= 5.698\ 585\ \dots \times 10^{-7}$$

Exercise Set 6.2.5

1.

$$3.6 \times 10^{3} \times 1.8 \times 10^{-4} + 1.9 \times 10^{1}$$
$$= 6.48 \times 10^{3-4} + 1.9 \times 10^{1}$$
$$= 6.48 \times 10^{-1} + 1.9 \times 10^{1}$$
$$= 0.0648 \times 10^{1} + 1.9 \times 10^{1}$$
$$= 1.9648 \times 10^{1}$$

2.

$$-7.4 \times 10^{62} \times 5.9 \times 10^{-33} - 2.4 \times 10^{28}$$
$$= -43.66 \times 10^{62-33} - 2.4 \times 10^{28}$$
$$= -43.66 \times 10^{29} - 2.4 \times 10^{28}$$
$$= -4.366 \times 10^{30} - 2.4 \times 10^{28}$$
$$= -4.366 \times 10^{30} - 0.024 \times 10^{30}$$
$$= -4.39 \times 10^{30}$$

3.

$$6.4 \times 10^{-42}\left(-8.7 \times 10^{-22}\right) + 3.2 \times 10^{-62}$$
$$= -55.68 \times 10^{-42-22} + 3.2 \times 10^{-62}$$
$$= -55.68 \times 10^{-64} + 3.2 \times 10^{-62}$$
$$= -0.5568 \times 10^{-62} + 3.2 \times 10^{-62}$$
$$= 2.6432 \times 10^{-62}$$

4.

$$2.9 \times 10^{-2} - \left(-1.7 \times 10^{-1}\right)\left(3.3 \times 10^{-2}\right)$$
$$= 2.9 \times 10^{-2} + 5.61 \times 10^{-1-2}$$
$$= 2.9 \times 10^{-2} + 5.61 \times 10^{-3}$$
$$= 2.9 \times 10^{-2} + 0.561 \times 10^{-2}$$
$$= 3.461 \times 10^{-2}$$

5.

$$7.1 \times 10^{17} - 8.81 \times 10^{42}\left(-6.2 \times 10^{-24}\right)$$
$$= 7.1 \times 10^{17} + 54.622 \times 10^{42-24}$$
$$= 7.1 \times 10^{17} + 54.622 \times 10^{18}$$
$$= 7.1 \times 10^{17} + 5.4622 \times 10^{19}$$
$$= 0.071 \times 10^{19} + 5.4622 \times 10^{19}$$
$$= 5.5332 \times 10^{19}$$

6.

$$-4.925 \times 10^{-8}\left(-3.7 \times 10^{0}\right) - 2.73 \times 10^{-17}\left(-4.5 \times 10^{10}\right)$$
$$= 18.2225 \times 10^{-8+0} + 12.285 \times 10^{-17+10}$$
$$= 18.2225 \times 10^{-8} + 12.285 \times 10^{-7}$$
$$= 1.822\ 25 \times 10^{-7} + 1.2285 \times 10^{-6}$$
$$= 0.182\ 225 \times 10^{-6} + 1.2285 \times 10^{-6}$$
$$= 1.410\ 725 \times 10^{-6}$$

7.

$$3.8 \times 10^{-72}\left(-1.5 \times 10^{29}\right) - 1.6 \times 10^{-80}\left(-1.1 \times 10^{37}\right)$$
$$= -5.7 \times 10^{-72+29} + 1.76 \times 10^{-80+37}$$
$$= -5.7 \times 10^{-43} + 1.76 \times 10^{-43}$$
$$= -3.94 \times 10^{-43}$$

8.

$$9.9 \times 10^{0}\left(2.1 \times 10^{-3}\right) - 5.7 \times 10^{-1}\left(8.9 \times 10^{-3}\right)$$
$$= 20.79 \times 10^{0-3} - 50.73 \times 10^{-1-3}$$
$$= 20.79 \times 10^{-3} - 50.73 \times 10^{-4}$$
$$= 2.079 \times 10^{-2} - 5.073 \times 10^{-3}$$
$$= 2.079 \times 10^{-2} - 0.5073 \times 10^{-2}$$
$$= 1.5717 \times 10^{-2}$$

9.

$$3 \times 10^{55} \times 4.1 \times 10^{-23} - 4.22 \times 10^{24} \times 2.6 \times 10^{8}$$
$$= 12.3 \times 10^{55-23} - 10.972 \times 10^{24+8}$$
$$= 12.3 \times 10^{32} - 10.972 \times 10^{32}$$
$$= 1.23 \times 10^{33} - 1.0972 \times 10^{33}$$
$$= 0.1328 \times 10^{33}$$
$$= 1.328 \times 10^{34}$$

10.

$$4.57 \times 10^{0} \left(-1.9 \times 10^{0}\right) - 3.4 \times 10^{2} \left(1.9 \times 10^{-2}\right)$$
$$= -8.683 \times 10^{0+0} - 6.46 \times 10^{2-2}$$
$$= -8.683 \times 10^{0} - 6.46 \times 10^{0}$$
$$= -15.143 \times 10^{0}$$
$$= -1.5143 \times 10^{1}$$

11.

$$3.4 \times 10^{5} \left(-2.1 \times 10^{3}\right) \left(9.2 \times 10^{-2}\right) - 1.8 \times 10^{7}$$
$$= -65.688 \times 10^{5+3-2} - 1.8 \times 10^{7}$$
$$= -65.688 \times 10^{6} - 1.8 \times 10^{7}$$
$$= -6.5688 \times 10^{7} - 1.8 \times 10^{7}$$
$$= -8.3688 \times 10^{7}$$

12.

$$9.3 \times 10^{-6} - 2.9 \times 10^{-2} \left(-3.2 \times 10^{-7}\right) \left(5.3 \times 10^{3}\right)$$
$$= 9.3 \times 10^{-6} + 49.184 \times 10^{-2-7+3}$$
$$= 9.3 \times 10^{-6} + 49.184 \times 10^{-6}$$
$$= 9.3 \times 10^{-6} + 4.9184 \times 10^{-5}$$
$$= 0.93 \times 10^{-5} + 4.9184 \times 10^{-5}$$
$$= 5.8484 \times 10^{-5}$$

13.

$$3.1 \times 10^{-2} \left(1.7 \times 10^{4} - 2.9 \times 10^{3}\right)$$
$$= 3.1 \times 10^{-2} \left(1.7 \times 10^{4} - 0.29 \times 10^{4}\right)$$
$$= 3.1 \times 10^{-2} \times 1.41 \times 10^{4}$$
$$= 4.371 \times 10^{-2+4}$$
$$= 4.371 \times 10^{2}$$

14.

$$-1.8 \times 10^3 \left(-5.2 \times 10^{-1} + 3.8 \times 10^{-2}\right)$$
$$= -1.8 \times 10^3 \left(-5.2 \times 10^{-1} + 0.38 \times 10^{-1}\right)$$
$$= -1.8 \times 10^3 \left(-4.82 \times 10^{-1}\right)$$
$$= 8.676 \times 10^{3-1}$$
$$= 8.676 \times 10^2$$

15.

$$4.2 \times 10^{-7} \left(2.1 \times 10^{-3} - 3.8 \times 10^{-3}\right)$$
$$= 4.2 \times 10^{-7} \left(-1.7 \times 10^{-3}\right)$$
$$= -7.14 \times 10^{-7-3}$$
$$= -7.14 \times 10^{-10}$$

16.

$$2.2 \times 10^0 \left(-6.4 \times 10^{-8} - 1.2 \times 10^{-7}\right)$$
$$= 2.2 \times 10^0 \left(-0.64 \times 10^{-7} - 1.2 \times 10^{-7}\right)$$
$$= 2.2 \times 10^0 \left(-1.84 \times 10^{-7}\right)$$
$$= -4.048 \times 10^{0-7}$$
$$= -4.048 \times 10^{-7}$$

17.

$$-1.8 \times 10^3 \left(2.9 \times 10^6 - 5.7 \times 10^7\right) + 2.4 \times 10^4 \left(3.2 \times 10^5 - 2.8 \times 10^4\right)$$
$$= -1.8 \times 10^3 \left(0.29 \times 10^7 - 5.7 \times 10^7\right) + 2.4 \times 10^4 \left(3.2 \times 10^5 - 0.28 \times 10^5\right)$$
$$= -1.8 \times 10^3 \left(-5.41 \times 10^7\right) + 2.4 \times 10^4 \left(2.92 \times 10^5\right)$$
$$= 9.738 \times 10^{3+7} + 7.008 \times 10^{4+5}$$
$$= 9.738 \times 10^{10} + 7.008 \times 10^9$$
$$= 9.738 \times 10^{10} + 0.7008 \times 10^{10}$$
$$= 10.4388 \times 10^{10}$$
$$= 1.04388 \times 10^{11}$$

18.

$$7.3 \times 10^{-8} \left(1.5 \times 10^0 + 2.2 \times 10^{-1}\right) - 1.8 \times 10^{-4} \left(9.2 \times 10^{-4} + 3.7 \times 10^{-3}\right)$$
$$= 7.3 \times 10^{-8} \left(1.5 \times 10^0 + 0.22 \times 10^0\right) - 1.8 \times 10^{-4} \left(0.92 \times 10^{-3} + 3.7 \times 10^{-3}\right)$$
$$= 7.3 \times 10^{-8} \left(1.72 \times 10^0\right) - 1.8 \times 10^{-4} \left(4.62 \times 10^{-3}\right)$$
$$= 12.556 \times 10^{-8+0} - 8.316 \times 10^{-4-3}$$
$$= 12.556 \times 10^{-8} - 8.316 \times 10^{-7}$$
$$= 1.2556 \times 10^{-7} - 8.316 \times 10^{-7}$$
$$= -7.0604 \times 10^{-7}$$

19.

$$1.9 \times 10^3 + \frac{3.7 \times 10^5}{5.7 \times 10^2}$$

$$= 1.9 \times 10^3 + 0.649\ 122\ \ldots \times 10^{5-2}$$

$$= 1.9 \times 10^3 + 0.649\ 122\ \ldots \times 10^3$$

$$= 2.549\ 122\ \ldots \times 10^3$$

20.

$$-6.3 \times 10^5 - \frac{1.8 \times 10^2}{-1.7 \times 10^{-2}}$$

$$= -6.3 \times 10^5 + 1.058\ 823\ \ldots \times 10^{2+2}$$

$$= -6.3 \times 10^5 + 1.058\ 823\ \ldots \times 10^4$$

$$= -6.3 \times 10^5 + 0.105\ 882\ \ldots \times 10^5$$

$$= -6.194\ 117\ \ldots \times 10^5$$

21.

$$\frac{1.2 \times 10^3}{2.5 \times 10^{-1}} - 2.7 \times 10^1$$

$$= 0.48 \times 10^{3+1} - 2.7 \times 10^1$$

$$= 0.48 \times 10^4 - 2.7 \times 10^1$$

$$= 4.8 \times 10^3 - 2.7 \times 10^1$$

$$= 4.8 \times 10^3 - 0.027 \times 10^3$$

$$= 4.773 \times 10^3$$

22.

$$\frac{-3.7 \times 10^{-3}}{-4.7 \times 10^2} - \left(-7.5 \times 10^{-4}\right)$$

$$= 0.787\ 234\ \ldots \times 10^{-3-2} + 7.5 \times 10^{-4}$$

$$= 0.787\ 234\ \ldots \times 10^{-5} + 7.5 \times 10^{-4}$$

$$= 7.872\ 340\ \ldots \times 10^{-6} + 7.5 \times 10^{-4}$$

$$= 0.078\ 723\ \ldots \times 10^{-4} + 7.5 \times 10^{-4}$$

$$= 7.578\ 723\ \ldots \times 10^{-4}$$

23.

$$\frac{5.5 \times 10^0}{-2.1 \times 10^3} + 4.3 \times 10^{-2}$$

$$= -2.619\ 047\ \ldots \times 10^{0-3} + 4.3 \times 10^{-2}$$

$$= -2.619\ 047\ \ldots \times 10^{-3} + 4.3 \times 10^{-2}$$

$$= -0.261\ 904\ \ldots \times 10^{-2} + 4.3 \times 10^{-2}$$

$$= 4.038\ 095\ \ldots \times 10^{-2}$$

Exercise Set 6.2.6

1. 5×10^2
2. 4.5×10^2
3. 4.52×10^2
4. 4.520×10^2
5. 4×10^0
6. 3.9×10^0
7. 3.89×10^0
8. 6×10^{-2}
9. -2×10^1
10. -1.7×10^1
11. -1.72×10^1
12. -1.720×10^1
13. -1×10^0
14. -8×10^{-1}
15. -7.7×10^{-1}
16. 4×10^3
17. 3.8×10^3
18. 3.82×10^3
19. 2×10^2
20. 1.9×10^2
21. 1.89×10^2
22. 0
23. 3×10^1
24. 0
25. 1×10^0
26. 1.0×10^0
27. 1.00×10^0
28. -8×10^3
29. -8.2×10^3
30. -8.24×10^3
31. 0
32. -2×10^1
33. -2.0×10^1
34. 1×10^1
35. 6×10^0
36. 6.3×10^0
37. 4×10^3
38. 3.8×10^3
39. 3.78×10^3
40. 1×10^0
41. 1.0×10^0
42. 1.00×10^0
43. 1.9×10^0
44. -8.4×10^{-2}
45. -1×10^0
46. -5×10^{-2}
47. 4.4×10^3
48. -8×10^{-3}
49. -2.0×10^1
50. 6.3×10^0
51. 2×10^2
52. 3×10^1
53. 4.3×10^2
54. 3.2×10^4
55. 0
56. 9.3×10^{-1}

Exercise Set 6.2.7

1. a. 5.25×10^4
   b. 3.300×10^{-3}
   c. -5.82×10^{-1}
   d. -8.10×10^0
   e. -7.206×10^{26}
   f. 6.17×10^{-51}
   g. 2.030×10^0
   h. -3.795×10^{-7}

2. a.

b.

c.

d.

e.

f.

g.

h.

i.

Exercise Set 7.1

1. A quantity is a property of an entity that can be measured objectively. An example of a property that is a quantity is *temperature*. An example of a property that is not a quantity is *beauty*.

2. A quantity name is a word used to refer to that quantity. An example of a quantity name is *temperature*.

3. A quantity symbol is a symbol used to represent the *value* of a quantity. An example of a quantity symbol is the symbol F used to represent the value of the quantity *force*.

4. Quantity symbols represent, not the quantities themselves, but their values. As such they can be used in mathematical expressions and equations. Quantity names, on the other hand, are words that are used to refer to the quantities themselves, not their values. As a result, it is not proper to use quantity names in mathematical expressions and equations.

Exercise Set 7.2

1. A unit is used to measure the value of a quantity. An example of a unit is the *metre* which is used to measure the value of the quantity *length*.

2. A unit name is a word used to refer to a unit. An example of a unit name is *metre*.

3. A unit symbol is a symbol used to represent the *size* of a unit. An example of a unit symbol is the symbol m used to represent the size of the unit *metre*.

$$-9.99 \quad\quad -9.98 \quad\quad -9.97 \quad\quad -9.96 \quad\quad -9.95 \quad\quad \times 10^{-18}$$

4. Unit symbols represent, not the units themselves, but their sizes. As such they can be used in mathematical expressions and equations. Unit names, on the other hand, are words that are used to refer to the units themselves, not their sizes. As a result, it is not proper to use unit names in mathematical expressions and equations.

Exercise Set 7.3

1. Quantity name: time
   Quantity symbol: t
   Entity: the test
   Value of quantity: 45 min
   Numerical value of quantity: 45
   Unit name: minute
   Unit symbol: min

2. Quantity name: length
   Quantity symbol: l
   Entity: beam
   Value of quantity: 5 ft
   Numerical value of quantity: 5
   Unit name: foot
   Unit symbol: ft

3. Quantity name: temperature
   Quantity symbol: T
   Entity: the solution
   Value of quantity: 285 K
   Numerical value of quantity: 285
   Unit name: kelvin
   Unit symbol: K

4. Quantity name: cost
   Quantity symbol: c
   Entity: the gloves
   Value of quantity: $25
   Numerical value of quantity: 25
   Unit name: dollar
   Unit symbol: $

5. Quantity name: mass
   Quantity symbol: m
   Entity: the medication
   Value of quantity: 10 mg
   Numerical value of quantity: 10
   Unit name: milligram
   Unit symbol: mg

6. Quantity name: time
   Quantity symbol: t
   Entity: my son
   Value of quantity: 4 years
   Numerical value of quantity: 4
   Unit name: year
   Unit symbol: year

7. Quantity name: time
   Quantity symbol: t
   Entity: the operation
   Value of quantity: 4.5 h
   Numerical value of quantity: 4.5
   Unit name: hour
   Unit symbol: h

8. Quantity name: volume
   Quantity symbol: V
   Entity: paint
   Value of quantity: 1.2 L
   Numerical value of quantity: 1.2
   Unit name: litre
   Unit symbol: L

9. Quantity name: speed
   Quantity symbol: s or v
   Entity: the car
   Value of quantity: 120 km/h
   Numerical value of quantity: 120
   Unit name: kilometres per hour
   Unit symbol: km/h

10. Quantity name: energy
    Quantity symbol: E
    Entity: the sandwich

Value of quantity: 240 Cal
Numerical value of quantity: 240
Unit name: Calorie
Unit symbol: Cal

11. Quantity name: distance
Quantity symbol: d
Entity: Montreal (to Toronto)
Value of quantity: 539 km
Numerical value of quantity: 539
Unit name: kilometre
Unit symbol: km

12. Quantity name: mass
Quantity symbol: m
Entity: CO_2
Value of quantity: 45.5 g
Numerical value of quantity: 45.5
Unit name: gram
Unit symbol: g

13. Quantity name: cost
Quantity symbol: c
Entity: the book
Value of quantity: $9.20
Numerical value of quantity: 9.20
Unit name: dollar
Unit symbol: $

14. Quantity name: rate
Quantity symbol: r
Entity: tax
Value of quantity:[3] 13%
Numerical value of quantity: 13%
Unit name: one
Unit symbol: 1

15. Quantity name: speed
Quantity symbol: s or v
Entity: the train
Value of quantity: 200 km/h
Numerical value of quantity: 200
Unit name: kilometres per hour
Unit symbol: km/h

16. Quantity name: area
Quantity symbol: A
Entity: the lot
Value of quantity: 4000 ft^2
Numerical value of quantity: 4000
Unit name: square feet
Unit symbol: ft^2

17. Quantity name: temperature
Quantity symbol: t
Entity: surface of the sun
Value of quantity: 6000 °C
Numerical value of quantity: 6000
Unit name: degree Celsius
Unit symbol: °C

18. Quantity name: energy
Quantity symbol: E
Entity: the bullet
Value of quantity: 2025 J
Numerical value of quantity: 2025
Unit name: joule
Unit symbol: J

19. Quantity name: acceleration
Quantity symbol: a
Entity: the BMW
Value of quantity: 12.66 m/s^2

---

[3]Formally, the value of this quantity is written as 13% 1, with 1 acting as the unit symbol for the unit of counting, *one*. The use of 1 as the unit of percentage arises from the interpretation of percentage as the ratio of two quantities with the same unit. As an example, when we speak of to 62% of the population, we are referring to 62 *people* out of every 100 *people*, i.e., a ratio of

$$\frac{62 \text{ people}}{100 \text{ people}}$$

or, more formally

$$\frac{62 \ 1}{100 \ 1}$$

Since division of two quantities with the same unit counts the number of times that the divisor fits into the dividend, the quotient will have the unit of counting.

In practice, however, the unit symbol, 1, is omitted, i.e., we write 13% in place of 13% 1.

Numerical value of quantity: 12.66   Unit symbol: kg
Unit name: metres per square second
Unit symbol: m/s^2

20. Quantity name: mass
    Quantity symbol: m
    Entity: Charlie
    Value of quantity: 32 kg
    Numerical value of quantity: 32
    Unit name: kilogram

21. Quantity name: weight
    Quantity symbol: W
    Entity: Charlie
    Value of quantity: 314 N
    Numerical value of quantity: 314
    Unit name: newton
    Unit symbol: N

Exercise Set 8.1

1. A measurement system consists of a list of quantities and their associated units along with conventions that establish a uniform communication protocol. An example of a measurement system is the International System of Units, SI.

2. The default measurement system used in the sciences is the International System of Units, SI.

3. The conventions provided by measurement systems aim to establish a uniform communication protocol.

4. In a given measurement system, a base quantity is a quantity that is given its own unit. An example of a base quantity in SI is length.

5. In a given measurement system, a base unit is a unit that is used to measure the value of a base quantity in that system. An example of a base unit in SI is *metre* which is used to measure the value of the SI base quantity *length*.

6. In a given measurement system, a derived quantity is a quantity whose unit is given in terms of that measurement system's base units. An example of a derived quantity in SI is speed.

7. In a given measurement system, a derived unit is a unit that is used to measure the value of a derived quantity in that system. An example of a derived unit in SI is *metres per second* which is used to measure the value of the SI derived quantity *speed*.

8. An example of a special name that is given to a combination of base units in SI is *joule* with corresponding unit symbol J. The symbol J represents the combination of SI base units kg·m^2/s^2.

9. Having a minimum number of base units in a measurement system simplifies the expression of science formulas within that system which would otherwise require the use of conversion factors to relate the values of the various quantities in that measurement system.

Exercise Set 8.2

1. a. There are seven base quantities in SI.
   b. The base quantity *length* is measured using the metre as the unit with unit symbol m. The base quantity mass is measured using the kilogram as the unit with unit symbol kg. The base quantity *time* is measured using the

second as the unit with unit symbol s. The base quantity *electric current* is measured using the ampere as the unit with unit symbol A. The base quantity *thermodynamic temperature* is measured using the kelvin as the unit with unit symbol K. The base quantity *amount of substance* is measured using the mole as the unit with unit symbol mol. The base quantity *luminous intensity* is measured using the candela as the unit with unit symbol cd.

2. a. An object's mass is a measure of that object's resistance to change in its velocity. An object's weight is a measure of force exerted on it by another object such as the Earth.

   b. The initial unit of mass, the gram, was too small for use in everyday life activities resulting in large numerical values accompanying the unit. It was replaced by the kilogram to bring the numerical values of masses that we encounter on a daily basis within a reasonable range.

3. The definition of the kelvin is based on the state of motion of molecules within a substance which in turn relates to the heat content of that substance. This establishes a relationship between temperature and heat that is more in line with physical behaviour of substances in contrast to other units of temperature which are defined in terms of behaviour of specific substances. The connection between the definition of the kelvin as the unit of temperature and heat means that the equations that relate these quantities are simpler in form and easier to comprehend.

4. a. The SI unit of area is the square metre.

   b. It means that the shape covers an area that is equivalent to the area covered by 2.4 unit squares.

   c. The phrase *square metre* is used to refer to an area as in 4 m^2 which is read as *four square metres*. The phrase *metre squared* refers to squaring a length as in $(4 \text{ m})^2$ which is short for 4 m $\times$ 4 m or 16 m^2. The expression $(4 \text{ m})^2$ is read as *four metres squared*.

5. a. The SI unit of volume is the cubic metre.

   b. It means that the object has a volume that is equivalent to the volume enclosed by 1.3 unit cubes.

   c. The phrase *cubic metre* is used to refer to a volume as in 4 m^3 which is read as *four cubic metres*. The phrase *metre cubed* refers to cubing a length as in $(4 \text{ m})^3$ which is short for 4 m $\times$ 4 m $\times$ 4 m or 64 m^3. The expression $(4 \text{ m})^3$ is read as *four metres cubed*.

6. a. The SI derived quantity *speed* is a measure of distance covered in unit time. The SI derived unit of speed is m/s.

   b. The SI derived unit of speed, i.e., m/s, relates to the SI base unit of length, i.e., m, and the SI base unit of time, i.e., s.

   c. The derived quantity *velocity* is defined as speed along with direction of motion.

7. a. The SI derived quantity *acceleration* is a measure of change in the velocity of an object in unit time. The SI derived unit of acceleration is m/s^2.

   b. The SI derived quantity *acceleration* is defined as change in speed in unit time, i.e.,

$$a = \frac{v}{t}$$

The SI derived unit of acceleration is, therefore, $(m/s)/s$. This can be simplified as follows:

$$(m/s)/s = \frac{\frac{m}{s}}{s}$$

$$= \frac{\frac{m}{s}}{\frac{s}{1}}$$

$$= \frac{m \times 1}{s \times s}$$

$$= \frac{m}{s^2}$$

$$= m/s^2$$

c. The SI derived unit of acceleration, i.e., m/s^2 which is short for $(m/s)/s$, relates to the SI base unit of length, i.e., m, and the SI base unit of time, i.e., s.

8. a. The SI derived quantity *density* is a measure of the mass of a substance in a unit volume of that substabce. The SI derived unit of density is kg/m^3.

b. The SI derived unit of density, i.e., kg/m^3, is related to the SI base unit of mass, i.e., kg, and the SI base unit of length, i.e., m.

9. a. The SI derived quantity *frequency* is a measure of the rate at which a cyclic process recurs. The SI derived unit of frequancy is hurtz with unit symbol Hz.

b. The SI derived unit of frequency, i.e., Hz which is short for $1/s$, is related to the SI unit of amount, i.e., 1, and the SI base unit of time, i.e., s.

10. a. The SI derived quantity *force* may be seen as a pull or a push. The SI derived unit of force is newton with unit symbol N.

b. The SI derived unit of force, i.e., N which is short for $kg{\cdot}m/s^2$, is related to the SI base unit of mass, i.e., kg, the SI base unit of length, i.e., m, and the SI base unit of time, i.e., s.

11. a. The SI derived quantity *pressure* is defined as force per unit area. The SI derived unit of pressure is pascal with unit symbol Pa.

b. The SI derived unit of pressure, i.e., Pa which is short for $kg/(m{\cdot}s^2)$, is related to the SI base unit of mass, i.e., kg, the SI base unit of length, i.e., m, and the SI base unit of time, i.e., s.

12. a. The SI derived quantity *work* is defined as force times distance. The SI derived unit of work is joule with unit symbol J.

b. The SI derived unit of work, i.e., J which is short for $kg{\cdot}m^2/s^2$, is related to the SI base unit of mass, i.e., kg, the SI base unit of length, i.e., m, and the SI base unit of time, i.e., s.

13. a. The SI derived quantity *energy* is defined as a quantity that can do work. The SI derived unit of energy is joule with unit symbol J.

b. Examples of different forms of energy are *kinetic energy* which is energy that an object possesses due to its state of motion, *sound energy* which is energy that is carried by sound waves, and *heat* which is the sum of the energies of molecules and atoms within a substance.

c. The SI derived unit of energy, i.e., J which is short for $kg \cdot m^2/s^2$, is related to the SI base unit of mass, i.e., kg, the SI base unit of length, i.e., m, and the SI base unit of time, i.e., s.

14. a. The SI derived quantity *power* is defined as energy used in unit time. The SI derived unit of power is watt with unit symbol W.

 b. The SI derived unit of power, i.e., W which is short for $kg \cdot m^2/s^3$, is related to the SI base unit of mass, i.e., kg, the SI base unit of length, i.e., m, and the SI base unit of time, i.e., s.

15. The commonly used non-SI units of time are minute, hour, day, week, month, year, decade, century and millennium.

16. a. The unit *atomic mass unit* was introduced into the sciences to make it more convenient to work with extremely small masses such as the masses of molecules, atoms and subatomic particles. The use of the SI unit of mass, the gram, in the expression of the values of the masses of such particles is inconvenient as its use would force the use of extrememly small numerical values.

 b. $1 \text{ g} = 6.023 \times 10^{23}$ amu

17. a. $1 \text{ L} = 1 \text{ dm}^3$

 b. $1 \text{ mL} = 1 \text{ cm}^3$

Exercise Set 8.3

1. The SI prefixes covered in this textbook along with their symbols and relative sizes to the unit are: kilo with symbol k which is 1000 times larger than the unit, hecto with symbol h which is 100 times larger than the unit, deca with symbol da which is 10 times larger than the unit, deci with symbol d which is 0.1 of the size of the unit, centi with symbol c which is 0.01 of the size of the unit, and milli with symbol m which is 0.001 of the size of the unit.

2. a. $1 \text{ km} = 1000 \text{ m}$ e. $1 \text{ km} = 10 \text{ hm}$ i. $1 \text{ cm} = 0.0001 \text{ hm}$
 b. $1 \text{ m} = 100 \text{ cm}$ f. $1 \text{ hm} = 1000 \text{ dm}$ j. $1 \text{ dam} = 0.01 \text{ km}$
 c. $1 \text{ m} = 1000 \text{ mm}$ g. $1 \text{ hm} = 10\,000 \text{ cm}$ k. $1 \text{ m} = 0.1 \text{ dam}$
 d. $1 \text{ cm} = 10 \text{ mm}$ h. $1 \text{ mm} = 0.01 \text{ dm}$ l. $1 \text{ dm} = 0.1 \text{ m}$

3. a. $1 \text{ kg} = 100 \text{ dag}$ d. $1 \text{ mN} = 0.1 \text{ cN}$ g. $1 \text{ cW} = 0.001 \text{ daW}$
 b. $1 \text{ mol} = 1000 \text{ mmol}$ e. $1 \text{ dacd} = 10 \text{ cd}$ h. $1 \text{ K} = 100 \text{ cK}$
 c. $1 \text{ A} = 100 \text{ cA}$ f. $1 \text{ hJ} = 1000 \text{ dJ}$ i. $1 \text{ cL} = 10 \text{ mL}$

4. a. $1 \text{ m}^2 = 100 \text{ dm}^2$ f. $1 \text{ hm}^3 = 1\,000\,000 \text{ m}^3$
 b. $1 \text{ m}^2 = 10\,000 \text{ cm}^2$ g. $1 \text{ dam}^3 = 1\,000\,000 \text{ dm}^3$
 c. $1 \text{ m}^2 = 1\,000\,000 \text{ mm}^2$ h. $1 \text{ dm}^3 = 0.000\,001 \text{ dam}^3$
 d. $1 \text{ cm}^2 = 0.0001 \text{ m}^2$ i. $1 \text{ cm}^3 = 1000 \text{ mm}^3$
 e. $1 \text{ hm}^2 = 0.01 \text{ km}^2$

5. a. $1\,\mathrm{s}^2 = 1\,000\,000\ \mathrm{ms}^2$
    b. $1\ \mathrm{K}^2 = 0.01\ \mathrm{daK}^2$
    c. $1\ \mathrm{mol}^2 = 100\ \mathrm{dmol}^2$
    d. $1\ \mathrm{cL}^2 = 0.01\ \mathrm{dL}^2$
    e. $1\ \mathrm{hg}^2 = 0.01\ \mathrm{kg}^2$

    f. $1\ \mathrm{hs}^3 = 1000\ \mathrm{das}^3$
    g. $1\ \mathrm{mmol}^3 = 0.001\ \mathrm{cmol}^3$
    h. $1\ \mathrm{mA}^3 = 0.000\,001\ \mathrm{dA}^3$
    i. $1\ \mathrm{daN}^3 = 1000\ \mathrm{N}^3$

6. a. 4250 cm
    b. 3610 mm
    c. 81 000 dam

    d. 0.35 mm
    e. 0.0241 m
    f. 4.25 km

    g. 3.1 hm
    h. 0.081 dam
    i. 0.001 41 km

7. a. 3800 cg
    b. 24 600 mL
    c. 3.54 hN

    d. 41 kJ
    e. 0.000 131 hPa
    f. 2.36 W

    g. 1.92 dK
    h. 0.0088 hA
    i. 0.135 damol

8. a. $1380\ \mathrm{dm}^2$
    b. $4\,250\,000\ \mathrm{mm}^2$
    c. $14\,000\,000\ \mathrm{dam}^2$

    d. $2490\ \mathrm{mm}^2$
    e. $0.000\,712\ \mathrm{hm}^2$
    f. $15\,200\ \mathrm{dm}^2$

    g. $0.448\ \mathrm{m}^2$
    h. $0.004\,73\ \mathrm{hm}^2$
    i. $0.001\,324\ \mathrm{hm}^2$

9. a. $37\,800\,000\ \mathrm{cm}^3$
    b. $612\,000\,000\ \mathrm{mm}^3$

    c. $4\,100\,000\ \mathrm{dam}^3$
    d. $5.2\ \mathrm{cm}^3$

    e. $0.000\,147\ \mathrm{m}^3$
    f. $0.000\,637\ \mathrm{km}^3$

10. a. $42\,500\,000\ \mathrm{ms}^2$
    b. $98\ \mathrm{dmol}^2$

    c. $0.0425\ \mathrm{km}^2$
    d. $0.247\ \mathrm{daK}^3$

    e. $0.001\,26\ \mathrm{cA}^3$
    f. $670\,000\ \mathrm{cJ}^3$

Exercise Set 8.4

1. SI conventions aim to enforce a uniform protocol for the communication of activities that pertain to the use of measured values in the sciences.

2. The italic font style should be used in the display of quantity symbols in SI. As an example, the quantity symbol for temperature is written as T, not T.

3. The casing of quantity symbols is important as different casings represent the values of different quantities. As an example, the quantity symbol P represents the value of the quantity *power* while the quantity symbol p represents the value of the quantity *pressure*.

4. a. m_{neutron} or m_{n}
    b. E_{kinetic} or E_{k}
    c. $m_{\mathrm{carbon\ dioxide}}$ or $m_{\mathrm{CO_2}}$
    d. V_{cube} or V_{c}
    e. T_{sun} or T_{s}

    f. p_{water} or p_{w}
    g. I_{wire} or I_{w}
    h. $E_{\mathrm{carbohydrates}}$ or E_{c}
    i. w_{lot} or w_{l}
    j. v_{car} or v_{c}

5. The upright font style should be used in the display of unit symbols in SI. As an example, the unit symbol for the metre is written as m, not m.

6. The casing of unit symbols is important as different casings represent the sizes of different units. As an example, the unit symbol mm represents the size of the unit *millimetre* while the unit symbol Mm represents the size of the unit *megametre*.

7. The convention to start unit names in SI with lowercase letters helps differentiate between the unit and the individual when the unit name is fashioned after the name of an individual. As an example, *newton* refers to the size of a unit while *Newton* refers to the individual.

8. The empty space between the numerical value of a quantity and the unit symbol that follows it implies multiplication. As an example, the expression 42 m represents 42×m.

9. Base-ten notations are easier to work with (compare, round, add, subtract, multiply, divide, etc.). In addition, the use of base-ten notations in the representation of numerical value of quantities makes it easier to convert between units in SI as such conversions force the movement of the decimal point in the numerical value of the quantities involved.

10. Multiplication of units in SI can be represnted as an empty space or the multiplication dot between the quantity symbols being multiplied. As an example, the product of the metre and the second can be written as either m s or m·s.

11. Division of units in SI can be represented as the horizontal line or the forward slash. An alternative notation is to use negative exponents on unit symbols that act as divisors. As an example, the division of metres by seconds can be written as $\frac{m}{s}$, m/s, m s^{-1}, or m·s^{-1}.

Exercise Set 9.1

1. a. ones or units   c. thousands   e. ones or units   g. ten thousandths
   b. tens   d. tenths   f. tenths   h. thousandths

2. a. Significant digits: 2, 4 and 7
   Estimated digit: 7

   b. Significant digits: 3, 5 and 8
   Estimated digit: 8

   c. Significant digits: 3, 1, 0 and 4
   Estimated digit: 4

   d. Significant digits: 5 and 6
   Estimated digit: 6

   e. Significant digits: 4 and 8
   Estimated digit: 8

   f. Significant digits: 3, 2 and 4
   Estimated digit: 4

   g. Significant digits: 1, 8 and 4
   Estimated digit: 4

   h. Significant digits: 2 and 9
   Estimated digit: 9

   i. Significant digits: 7, 0 and 2
   Estimated digit: 2

   j. Significant digits: 4, 1, 2 and 0
   Estimated digit: 0

   k. Significant digits: 3, 2, 1 and 8
   Estimated digit: 8

   l. Significant digits: 2, 0 and 0
   Estimated digit: Rightmost 0

   m. Significant digits: 2, 4 and 3
   Estimated digit: 3

   n. Significant digits: 7, 2 and leftmost 0
   Estimated digit: Leftmost 0

   o. Significant digits: 9, 4, 2 and 0
   Estimated digit: 0

   p. Significant digits: 3, two leftmost 0s
   Estimated digit: Middle 0

   q. Significant digits: 3 and 2
   Estimated digit: 2

   r. Significant digits: 3 and 2

Estimated digit: 2

s. Significant digits: 1, 4 and 7
   Estimated digit: 7

t. Significant digits: 2, 0 and 0
   Estimated digit: Rightmost 0

3. Amount

4. The defined value 100 in 1 m = 100 cm

5. a. 1 and 2   b. 2   c. 1, 2 and 2   d. 2

Exercise Set 9.2

1. a. $T \in (\,32.8, 38.7\,)$
   b. $m \in (\,2.5, 3.7\,]$
   c. $E \in [\,14\,000, 15\,200\,)$
   d. $A \in [\,95.4, 102.3\,]$

   e. $h \in (\,162, 178\,)$
   f. $g \in (\,52\%, 87\%\,]$
   g. $a \in [\,21.72, 23.80\,)$
   h. $v \in [\,90, 120\,]$

2. a.

$$318 + 27 + 140 = 485$$

Exact value: 485
Standard rule: 490
Three-significant-digit-rule: 485
Four-significant-digit-rule: 485.0
Five-significant-digit-rule: 485.00

   b.

$$52 + 33 - 123 + 43 = 5$$

Exact value: 5
Standard rule: 5
Three-significant-digit-rule: 5.00
Four-significant-digit-rule: 5.000
Five-significant-digit-rule: 5.0000

   c.

$$130 - 240 + 410 = 300$$

Exact value: 300
Standard rule: 300
Three-significant-digit-rule: 300
Four-significant-digit-rule: 300.0
Five-significant-digit-rule: 300.00

d.

$$-142 - 210 - 128 = -480$$

Exact value: -480
Standard rule: -480
Three-significant-digit-rule: -480
Four-significant-digit-rule: -480.0
Five-significant-digit-rule: -480.00

e.

$$4.25 + 1.86 - 2.1 = 4.01$$

Exact value: 4.01
Standard rule: 4.0
Three-significant-digit-rule: 4.01
Four-significant-digit-rule: 4.010
Five-significant-digit-rule: 4.0100

f.

$$3.8 + 4.51 - 5.20 = 3.11$$

Exact value: 3.11
Standard rule: 3.1
Three-significant-digit-rule: 3.11
Four-significant-digit-rule: 3.110
Five-significant-digit-rule: 3.1100

g.

$$9.73 + 9.8 - 6.542 = 12.988$$

Exact value: 12.988
Standard rule: 13.0
Three-significant-digit-rule: 13.0
Four-significant-digit-rule: 12.99
Five-significant-digit-rule: 12.988

h.

$$18.25 + 19 - 22.3 = 14.95$$

Exact value: 14.95
Standard rule: 15
Three-significant-digit-rule: 15.0
Four-significant-digit-rule: 14.95
Five-significant-digit-rule: 14.950

i.

$$1800 + 1920 - 3110 = 610$$

Exact value: 610
Standard rule: 600
Three-significant-digit-rule: 610
Four-significant-digit-rule: 610.0
Five-significant-digit-rule: 610.00

j.

$$4524 + 1830 - 9720 = -3366$$

Exact value: −3366
Standard rule: −3400
Three-significant-digit-rule: −3370
Four-significant-digit-rule: −3366
Five-significant-digit-rule: −3366.0

k.

$$0.0718 - 0.023\,47 - 0.0873 = -0.038\,97$$

Exact value: −0.038 97
Standard rule: −0.04
Three-significant-digit-rule: −0.0390
Four-significant-digit-rule: −0.038 97
Five-significant-digit-rule: −0.038 970

l.

$$4.5 \times 10^3 + 1.7 \times 10^3 = 6.2 \times 10^3$$

Exact value: 6.2×10^3
Standard rule: 6.2×10^3
Three-significant-digit-rule: 6.20×10^3
Four-significant-digit-rule: 6.200×10^3
Five-significant-digit-rule: 6.2000×10^3

m.

$$-1.9 \times 10^{-2} - 9.72 \times 10^{-2} = -11.62 \times 10^{-2}$$
$$= -1.162 \times 10^{-1}$$

Exact value: -1.162×10^{-1}
Standard rule: -1.16×10^{-1}
Three-significant-digit-rule: -1.16×10^{-1}
Four-significant-digit-rule: -1.162×10^{-1}
Five-significant-digit-rule: -1.1620×10^{-1}

n.

$$7.6 \times 10^2 + 8.8 \times 10^3 = 0.76 \times 10^3 + 8.8 \times 10^3$$
$$= 9.\underset{.}{5}6 \times 10^3$$

Exact value: 9.56×10^3
Standard rule: 9.6×10^3
Three-significant-digit-rule: 9.56×10^3
Four-significant-digit-rule: 9.560×10^3
Five-significant-digit-rule: 9.5600×10^3

o.

$$7.4 \times 10^0 + 1.8 \times 10^{-1} = 7.4 \times 10^0 + 0.18 \times 10^0$$
$$= 7.\underset{.}{5}8 \times 10^0$$

Exact value: 7.58×10^0
Standard rule: 7.6×10^0
Three-significant-digit-rule: 7.58×10^0
Four-significant-digit-rule: 7.580×10^0
Five-significant-digit-rule: 7.5800×10^0

p.

$$8.5 \times 10^{-3} + 6.7 \times 10^{-4} = 8.5 \times 10^{-3} + 0.67 \times 10^{-3}$$
$$= 9.\underset{.}{1}7 \times 10^{-3}$$

Exact value: 9.17×10^{-3}
Standard rule: 9.2×10^{-3}
Three-significant-digit-rule: 9.17×10^{-3}
Four-significant-digit-rule: 9.170×10^{-3}
Five-significant-digit-rule: 9.1700×10^{-3}

q.

$$24\,3\underset{.}{0}0 + 44\,0\underset{.}{0}0 - 57\,00\underset{.}{0} = 11\,3\underset{.}{0}0$$

Exact value: $11\,300$
Standard rule: $11\,300$
Three-significant-digit-rule: $11\,300$
Four-significant-digit-rule: $11\,3\underset{.}{0}0$
Five-significant-digit-rule: $11\,30\underset{.}{0}$

r.

$$3000 - 2560 + 1020 - 180\underset{.}{0} = -3\underset{.}{4}0$$

Exact value: -340
Standard rule: -300
Three-significant-digit-rule: $-34\underset{.}{0}$
Four-significant-digit-rule: -340.0
Five-significant-digit-rule: -340.00

s.

$$0.026 - 0.001\,82 + 0.0513 = 0.075\,\dot{4}8$$

Exact value: 0.075 48
Standard rule: 0.0755
Three-significant-digit-rule: 0.0755
Four-significant-digit-rule: 0.075 48
Five-significant-digit-rule: 0.075 480

t.

$$50\,200 + 32\,000 - 1010 + 21\,000 = 10\dot{2}\,190$$

Exact value: 102 190
Standard rule: 102 000
Three-significant-digit-rule: 102 000
Four-significant-digit-rule: 102 200
Five-significant-digit-rule: 102 190

3. a.

$$4.2 \times 3.71 = 1\dot{5}.582$$

Exact value: 15.582
Standard rule: 16
Three-significant-digit-rule: 15.6
Four-significant-digit-rule: 15.58
Five-significant-digit-rule: 15.582

b.

$$3.6 \times 2.45 \times 9.8 = 8\dot{6}.436$$

Exact value: 86.436
Standard rule: 86
Three-significant-digit-rule: 86.4
Four-significant-digit-rule: 86.44
Five-significant-digit-rule: 86.436

c.

$$0.025 \times 30\dot{0} = 7.\dot{5}$$

Exact value: 7.5
Standard rule: 7.5
Three-significant-digit-rule: 7.50
Four-significant-digit-rule: 7.500
Five-significant-digit-rule: 7.5000

d.

$$1902 \times 420\dot{0} = 7\,9\dot{8}8\,400$$

Exact value: 7 988 400
Standard rule: 7 990 000
Three-significant-digit-rule: 7 990 000
Four-significant-digit-rule: 7 988 000
Five-significant-digit-rule: 7 988 400

e.

$$0.001\,60 \times 240 \; = \; 0.384$$

Exact value: 0.384
Standard rule: 0.384
Three-significant-digit-rule: 0.384
Four-significant-digit-rule: 0.3840
Five-significant-digit-rule: 0.384 00

f.

$$0046 \times 205 \; = \; 9430$$

Exact value: 9430
Standard rule: 9400
Three-significant-digit-rule: 9430
Four-significant-digit-rule: 9430
Five-significant-digit-rule: 9430.0

g.

$$32.5 \times 8.4 \; = \; 273$$

Exact value: 273
Standard rule: 270
Three-significant-digit-rule: 273
Four-significant-digit-rule: 273.0
Five-significant-digit-rule: 273.00

h.

$$10.201 \times 42 \times 1200 \; = \; 514\,130.4$$

Exact value: 514 130.4
Standard rule: 510 000
Three-significant-digit-rule: 514 000
Four-significant-digit-rule: 514 100
Five-significant-digit-rule: 514 130

i.

$$\frac{62.5}{18} \; = \; 3.472\,222\ldots$$

Exact value: 3.472 222 . . .
Standard rule: 3.5

Three-significant-digit-rule: 3.47
Four-significant-digit-rule: 3.472
Five-significant-digit-rule: 3.4722

j.

$$\frac{9.6}{4.25} \times 3.60 = 8.131\,764\ldots$$

Exact value: 8.131 764 . . .
Standard rule: 8.1
Three-significant-digit-rule: 8.13
Four-significant-digit-rule: 8.132
Five-significant-digit-rule: 8.1318

k.

$$\frac{2400}{30\,000} \times \frac{7.69}{1.78 \times 21.9} = 0.015\,781\,642\,809\ldots$$

Exact value: 0.015 781 642 809 . . .
Standard rule: 0.016
Three-significant-digit-rule: 0.0158
Four-significant-digit-rule: 0.015 78
Five-significant-digit-rule: 0.015 782

l.

$$\frac{2.681 \times 4.5}{6.23} = 1.936\,516\ldots$$

Exact value: 1.936 516 . . .
Standard rule: 1.9
Three-significant-digit-rule: 1.94
Four-significant-digit-rule: 1.937
Five-significant-digit-rule: 1.9365

m.

$$\frac{1200}{420 \times 100} = 0.028\,571\,428\ldots$$

Exact value: 0.028 571 428 . . .
Standard rule: 0.029
Three-significant-digit-rule: 0.0286
Four-significant-digit-rule: 0.028 57
Five-significant-digit-rule: 0.028 571

n.

$$\frac{132\,000}{4200 \times 10.0} = 3.142\,857\ldots$$

Exact value: 3.142 857 . . .
Standard rule: 3.14

Three-significant-digit-rule: 3.14
Four-significant-digit-rule: 3.143
Five-significant-digit-rule: 3.1429

o.

$$4.7 \times 10^4 \left(3.2 \times 10^3\right) = 15.04 \times 10^{4+3}$$
$$= 15.04 \times 10^7$$
$$= 1.504 \times 10^8$$

Exact value: 1.504×10^8
Standard rule: 1.5×10^8
Three-significant-digit-rule: 1.50×10^8
Four-significant-digit-rule: 1.504×10^8
Five-significant-digit-rule: 1.5040×10^8

p.

$$-1.5 \times 10^{-3} \left(6.62 \times 10^5\right) = -9.93 \times 10^{-3+5}$$
$$= -9.93 \times 10^2$$

Exact value: -9.93×10^2
Standard rule: -9.9×10^2
Three-significant-digit-rule: -9.93×10^2
Four-significant-digit-rule: -9.930×10^2
Five-significant-digit-rule: -9.9300×10^2

q.

$$-3.7 \times 10^0 \left(2.5 \times 10^{-2}\right) \left(-8.4 \times 10^4\right) = 77.7 \times 10^{0-2+4}$$
$$= 77.7 \times 10^2$$
$$= 7.77 \times 10^3$$

Exact value: 7.77×10^3
Standard rule: 7.8×10^3
Three-significant-digit-rule: 7.77×10^3
Four-significant-digit-rule: 7.770×10^3
Five-significant-digit-rule: 7.7700×10^3

r.

$$\frac{2.571 \times 10^{-7} \left(1.70 \times 10^3\right)}{-3.6 \times 10^{-1}} = -1.214\,083\ldots \times 10^{-7+3+1}$$
$$= -1.214\,083\ldots \times 10^{-3}$$

Exact value: $-1.214\,083\ldots \times 10^{-3}$
Standard rule: -1.2×10^{-3}
Three-significant-digit-rule: -1.21×10^{-3}
Four-significant-digit-rule: -1.214×10^{-3}
Five-significant-digit-rule: -1.2141×10^{-3}

s.

$$\frac{-4.25 \times 10^{-4}}{5.72 \times 10^3 \left(-1.1 \times 10^{-2}\right)} = 0.675\,460\ldots \times 10^{-4-3+2}$$
$$= 0.675\,460\ldots \times 10^{-5}$$
$$= 6.754\,609\ldots \times 10^{-6}$$

Exact value: $6.754\,609\ldots \times 10^{-6}$
Standard rule: 6.8×10^{-6}
Three-significant-digit-rule: 6.75×10^{-6}
Four-significant-digit-rule: 6.755×10^{-6}
Five-significant-digit-rule: 6.7546×10^{-6}

t.

$$1.6 \times 3.62 \times 1.50 = 8.688$$

Exact value: 8.688
Standard rule: 8.69
Three-significant-digit-rule: 8.69
Four-significant-digit-rule: 8.688
Five-significant-digit-rule: 8.6880

u.

$$\frac{9.6}{4.25} = 2.258\,823\ldots$$

Exact value: $2.258\,823\ldots$
Standard rule: 2.26
Three-significant-digit-rule: 2.26
Four-significant-digit-rule: 2.259
Five-significant-digit-rule: 2.2588

v.

$$\frac{9.6}{4.25} = 2.258\,823\ldots$$

Exact value: $2.258\,823\ldots$
Standard rule: 2.3
Three-significant-digit-rule: 2.26
Four-significant-digit-rule: 2.259
Five-significant-digit-rule: 2.2588

w.

$$3.6 \times 60 \times 60 = 12\,960$$

Exact value: $12\,960$
Standard rule: $13\,000$
Three-significant-digit-rule: $13\,000$
Four-significant-digit-rule: $12\,960$
Five-significant-digit-rule: $12\,960$

x.

$$\frac{3.6 \times 400}{3.5 \times 1.82} = 226.059\,654\ldots$$

Exact value: 226.059 654...
Standard rule: 226
Three-significant-digit-rule: 226
Four-significant-digit-rule: 226.1
Five-significant-digit-rule: 226.06

4. a.

$$340 \times 24 + 9.8 \times 1200 = 8160 + 11\,760$$
$$= 19\,920$$

Exact value: 19 920
Standard rule: 20 000
Three-significant-digit-rule: 19 900
Four-significant-digit-rule: 19 920
Five-significant-digit-rule: 19 920

b.

$$72 \times 130 - 95 \times 210 + 28 \times 150 = 9360 - 19\,950 + 4200$$
$$= -6390$$

Exact value: −6390
Standard rule: −6000
Three-significant-digit-rule: −6390
Four-significant-digit-rule: −6390
Five-significant-digit-rule: −6390.0

c.

$$42 \times 120 \times 37 - 27 \times 1400 - 1400 \times 2.8 = 186\,480 - 37\,800 - 3920$$
$$= 144\,760$$

Exact value: 144 760
Standard rule: 140 000
Three-significant-digit-rule: 145 000
Four-significant-digit-rule: 144 800
Five-significant-digit-rule: 144 760

d.

$$4.2 \times 6.7 + 18 \times 1.2 = 28.14 + 21.6$$
$$= 49.74$$

Exact value: 49.74
Standard rule: 50
Three-significant-digit-rule: 49.7
Four-significant-digit-rule: 49.74
Five-significant-digit-rule: 49.740

e.

$$3.0 \times 42.5 + 9.68 \times 2.1 \times 1.04 = 127.5 + 21.141\,12$$
$$= 148.641\,12$$

Exact value: 148.641 12
Standard rule: 150
Three-significant-digit-rule: 150
Four-significant-digit-rule: 148.6
Five-significant-digit-rule: 148.64

f.

$$1.6 + 12.4 \times 1.72 + 0.094 = 1.6 + 21.328 + 0.094$$
$$= 23.022$$

Exact value: 23.022
Standard rule: 23.0
Three-significant-digit-rule: 23.0
Four-significant-digit-rule: 23.02
Five-significant-digit-rule: 23.022

g.

$$\frac{4.17 \times 1.5}{3.61} + 1.9 \times 4.2 - \frac{1.38}{3.6 \times 1.5}$$
$$= 1.732\,686\ldots + 7.98 - 0.255\,555\ldots$$
$$= 9.457\,130\ldots$$

Exact value: 9.457 130 . . .
Standard rule: 9.5
Three-significant-digit-rule: 9.46
Four-significant-digit-rule: 9.457
Five-significant-digit-rule: 9.4571

h.

$$\frac{4.56}{3.6} + \frac{18}{10.0} = 1.266\,666\ldots + 1.8$$
$$= 3.066\,666\ldots$$

Exact value: 3.066 666 . . .
Standard rule: 3.1
Three-significant-digit-rule: 3.07
Four-significant-digit-rule: 3.067
Five-significant-digit-rule: 3.0667

i.

$$0.87 \times 3.4 + 4.50 \times 0.652 = 2.958 + 2.934$$
$$= 5.892$$

Exact value: 5.892
Standard rule: 5.9

Three-significant-digit-rule: 5.89
Four-significant-digit-rule: 5.892
Five-significant-digit-rule: 5.8920

j.

$$14\,200 + 16 \times 2000 - 56\,000 = 14\,200 + 32\,000 - 56\,000$$
$$= -9800$$

Exact value: -9800
Standard rule: $-10\,000$
Three-significant-digit-rule: -9800
Four-significant-digit-rule: -9800
Five-significant-digit-rule: -9800.0

k.

$$2450 + 3000 \times 24 \times \frac{1}{650} = 2450 + 110.769\,230\ldots$$
$$= 2560.769\,230\ldots$$

Exact value: $2560.769\,230\ldots$
Standard rule: 2560
Three-significant-digit-rule: 2560
Four-significant-digit-rule: 2561
Five-significant-digit-rule: 2560.8

l.

$$2.57 \times 10^3 \left(1.8 \times 10^2\right) + 3.5 \times 10^2 \left(-4.7 \times 10^4\right)$$
$$= 4.626 \times 10^{3+2} - 16.45 \times 10^{2+4}\ldots$$
$$= 4.626 \times 10^5 - 16.45 \times 10^6$$
$$= 4.626 \times 10^5 - 1.645 \times 10^7$$
$$= 0.046\,26 \times 10^7 - 1.645 \times 10^7$$
$$= -1.598\,74 \times 10^7$$

Exact value: $-1.598\,74 \times 10^7$
Standard rule: -1.6×10^7
Three-significant-digit-rule: -1.60×10^7
Four-significant-digit-rule: -1.599×10^7
Five-significant-digit-rule: -1.5987×10^7

m.

$$-3.2 \times 10^0 \left(2.8 \times 10^{-3}\right) - 8.2 \times 10^{-1} \left(7.84 \times 10^{-1}\right)$$
$$= -8.96 \times 10^{0-3} - 64.288 \times 10^{-1-1}$$
$$= -8.96 \times 10^{-3} - 64.288 \times 10^{-2}$$
$$= -8.96 \times 10^{-3} - 6.4288 \times 10^{-1}$$
$$= -0.0896 \times 10^{-1} - 6.4288 \times 10^{-1}$$
$$= -6.5184 \times 10^{-1}$$

Exact value: -6.5184×10^{-1}
Standard rule: -6.5×10^{-1}

Three-significant-digit-rule: -6.52×10^{-1}
Four-significant-digit-rule: -6.518×10^{-1}
Five-significant-digit-rule: -6.5184×10^{-1}

n.

$$-1.99 \times 10^{-2} \left(-4.76 \times 10^{-4}\right) \left(1.1 \times 10^{4}\right) + 8.6 \times 10^{-3}$$
$$= 10.419\,64 \times 10^{-2-4+4} + 8.6 \times 10^{-3}$$
$$= 10.419\,64 \times 10^{-2} + 8.6 \times 10^{-3}$$
$$= 1.041\,964 \times 10^{-1} + 8.6 \times 10^{-3}$$
$$= 1.041\,964 \times 10^{-1} + 0.086 \times 10^{-1}$$
$$= 1.127\,964 \times 10^{-1}$$

Exact value: $1.127\,964 \times 10^{-1}$
Standard rule: 1.1×10^{-1}
Three-significant-digit-rule: 1.13×10^{-1}
Four-significant-digit-rule: 1.128×10^{-1}
Five-significant-digit-rule: 1.1280×10^{-1}

o.

$$\frac{1.8 \times 10^{2}}{-3.22 \times 10^{-1}} - \frac{6.58 \times 10^{-2}}{-1.52 \times 10^{-3}}$$
$$= -0.559\,006\,21 \ldots \times 10^{2+1} + 4.328\,947 \ldots \times 10^{-2+3}$$
$$= -0.559\,006\,21 \ldots \times 10^{3} + 4.328\,947 \ldots \times 10^{1}$$
$$= -5.590\,062 \ldots \times 10^{2} + 4.328\,947 \ldots \times 10^{1}$$
$$= -5.590\,062 \ldots \times 10^{2} + 0.432\,894 \ldots \times 10^{2}$$
$$= -5.157\,168 \ldots \times 10^{2}$$

Exact value: $-5.157\,168 \ldots \times 10^{2}$
Standard rule: -5.2×10^{2}
Three-significant-digit-rule: -5.16×10^{2}
Four-significant-digit-rule: -5.157×10^{2}
Five-significant-digit-rule: -5.1572×10^{2}

p.

$$4 \times 9.4 + 3 \times 0.64 - 2 \times 14.5 = 37.6 + 1.92 - 29.0$$
$$= 10.52$$

Exact value: 10.52
Standard rule: 11
Three-significant-digit-rule: 10.5
Four-significant-digit-rule: 10.52
Five-significant-digit-rule: 10.520

q.

$$-6 \times 3.7 \times 2.1 - (-2 \times 5.5) = -46.62 + 11$$
$$= -35.62$$

Exact value: -35.62
Standard rule: -36
Three-significant-digit-rule: -35.6
Four-significant-digit-rule: -35.62
Five-significant-digit-rule: -35.620

Exercise Set 10.1

1. It means that performing the operations on the numerical values and quantity symbols results in a value.

2. a. The sequence qualifies as an expression as it does not contain the equality or inequality symbols and is meaningful, i.e., it evaluates to a value: It requires that we multiply 2 by p and then add 3 to the result.

   b. The sequence qualifies as an expression as it does not contain the equality or inequality symbols and is meaningful, i.e., it evaluates to a value: It requires that we start with m, subtract 2, and then add the product of 3 and m to the result.

   c. The sequence does not qualify as an expression as it contains the equality symbol.

   d. The sequence qualifies as an expression as it does not contain the equality or inequality symbols and is meaningful, i.e., it evaluates to a value: It requires that we subtract the product of 6 and E from 14.

   e. The sequence qualifies as an expression as it does not contain the equality or inequality symbols and is meaningful, i.e., it evaluates to a value: It requires that we add the product of -5 and t to the product of 3 and v.

   f. The sequence does not qualify as an expression as it is meaningless: The sequence $- \div \times$ does not make sense.

   g. The sequence qualifies as an expression as it does not contain the equality or inequality symbols and is meaningful, i.e., it evaluates to a value: -8.

   h. The sequence qualifies as an expression as it does not contain the equality or inequality symbols and is meaningful, i.e., it evaluates to a value: A.

   i. The sequence does not qualify as an expression as it does not evaluate to a value.

   j. The sequence does not qualify as an expression as it is meaningless: It is not quite clear what it is that we intend to add to 8.

   k. The sequence qualifies as an expression as it does not contain the equality or inequality symbols and is meaningful, i.e., it evaluates to a value: It requires that we subtract 1 from the square of t.

   l. The sequence does not qualify as an expression as it contains the equality symbol.

   m. The sequence qualifies as an expression as it does not contain the equality or inequality symbols and is meaningful, i.e., it evaluates to a value: $\frac{3}{5}$.

   n. The sequence does not qualify as an expression as it contains an inequality symbol.

o. The sequence qualifies as an expression as it does not contain the equality or inequality symbols and is meaningful, i.e., it evaluates to a value: It requires that we multiply -3 by the sum of the product of 2 and p and 1.

3. a. The main structural components of the equation $-t + 7 = 2$ are the expressions $-t + 7$ and 2.

b. The main structural components of the equation $5 - r = 0$ are the expressions $5 - r$ and 0.

c. The main structural components of the equation $-3n + 2 = 5 + 2n$ are the expressions $-3n + 2$ and $5 + 2n$.

d. The main structural components of the equation $4(p - 1) = -3$ are the expressions $4(p - 1)$ and -3.

e. The main structural components of the equation $\frac{2}{3}T = \frac{1}{2}T + \frac{3}{5}$ are the expressions $\frac{2}{3}T$ and $\frac{1}{2}T + \frac{3}{5}$.

f. The main structural components of the equation $5A + 2 = -3(2A - 1)$ are the expressions $5A + 2$ and $-3(2A - 1)$.

g. The main structural components of the equation $2r - 3s = 5$ are the expressions $2r - 3s$ and 5.

h. The main structural components of the equation $t^2 + 2t + 3 = 0$ are the expressions $t^2 + 2t + 3$ and 0.

4. a. $r = 1$ is a solution to the equation $3r + 2 = 5$ as substitution of 1 for r in the equation results in equal values for the expressions on the left and right sides of the equation:

$$\begin{aligned} \text{LS} &= 3r + 2 \\ &= 3 \times 1 + 2 \\ &= 3 + 2 \\ &= 5 \end{aligned}$$

and

$$\text{RS} = 5$$

b. $t = 3$ is not a solution to the equation $-t - 4 = 0$ as substitution of 3 for t in the equation does not result in equal values for the expressions on the left and right sides of the equation:

$$\begin{aligned} \text{LS} &= -t - 4 \\ &= -3 - 4 \\ &= -7 \end{aligned}$$

but

$$\text{RS} = 0$$

c. $n = -3$ is a solution to the equation $2n + 3 = -6 - n$ as substitution of -3 for n in the equation results in equal values for the expressions on the left and right sides of the equation:

$$\begin{aligned} \text{LS} &= 2n + 3 \\ &= 2(-3) + 3 \\ &= -6 + 3 \\ &= -3 \end{aligned}$$

and

$$\begin{aligned} \text{RS} &= -6 - n \\ &= -6 - (-3) \\ &= -6 + 3 \\ &= -3 \end{aligned}$$

d. $p = 1$ is a solution to the equation $7 - 5p = 2p$ as substitution of 1 for p in the equation results in equal values for the expressions on the left and right sides of the equation:

$$\begin{aligned} \text{LS} &= 7 - 5p \\ &= 7 - 5 \times 1 \\ &= 7 - 5 \\ &= 2 \end{aligned}$$

and

$$\begin{aligned} \text{RS} &= 2p \\ &= 2 \times 1 \\ &= 2 \end{aligned}$$

e. $r = -1$ is not a solution to the equation $-(r - 1) + 2 = 0$ as substitution of -1 for r in the equation does not result in equal values for the expressions on the left and right sides of the equation:

$$\begin{aligned} \text{LS} &= -(r - 1) + 2 \\ &= -(-1 - 1) + 2 \\ &= -(-2) + 2 \\ &= 2 + 2 \\ &= 4 \end{aligned}$$

but

$$\text{RS} = 0$$

f. $v = 0$ is not a solution to the equation $3(-2v + 3) = -2 + 7v$ as substitution of 0 for v in the equation does not result in equal values for the expressions on the left and right sides of the equation:

$$\begin{aligned} \text{LS} &= 3\,(-2v\ +\ 3) \\ &= 3\,(-2 \times 0\ +\ 3) \\ &= 3\,(0\ +\ 3) \\ &= 3 \times 3 \\ &= 9 \end{aligned}$$

but

$$\begin{aligned} \text{RS} &= -2\ +\ 7v \\ &= -2\ +\ 7 \times 0 \\ &= -2\ +\ 0 \\ &= -2 \end{aligned}$$

g. $L = 3$ is not a solution to the equation $L^2 - 3L + 2 = 0$ as substitution of 3 for L in the equation does not result in equal values for the expressions on the left and right sides of the equation:

$$\begin{aligned} \text{LS} &= L^2\ -\ 3L\ +\ 2 \\ &= 3^2\ -\ 3 \times 3\ +\ 2 \\ &= 9\ -\ 9\ +\ 2 \\ &= 2 \end{aligned}$$

but

$$\text{RS} = 0$$

h. $m = -1$ is a solution to the equation $2m^2 = 6m + 8$ as substitution of -1 for m in the equation results in equal values for the expressions on the left and right sides of the equation:

$$\begin{aligned} \text{LS} &= 2m^2 \\ &= 2\,(-1)^2 \\ &= 2 \times 1 \\ &= 2 \end{aligned}$$

and

$$\begin{aligned} \text{RS} &= 6m\ +\ 8 \\ &= 6\,(-1)\ +\ 8 \\ &= -6\ +\ 8 \\ &= 2 \end{aligned}$$

i. $E = \frac{14}{15}$ is a solution to the equation $-\frac{1}{2}E + \frac{2}{3} = \frac{3}{4}E - \frac{1}{2}$ as substitution of $\frac{14}{15}$ for E in the equation results in equal values for the expressions on the left and right sides of the equation:

$$LS = -\frac{1}{2}E + \frac{2}{3}$$

$$= -\frac{1}{2} \times \frac{14}{15} + \frac{2}{3}$$

$$= -\frac{7}{15} + \frac{2}{3}$$

$$= \frac{-7 + 10}{15}$$

$$= \frac{3}{15}$$

$$= \frac{1}{5}$$

and

$$RS = \frac{3}{4}E - \frac{1}{2}$$

$$= \frac{3}{4} \times \frac{14}{15} - \frac{1}{2}$$

$$= \frac{7}{10} - \frac{1}{2}$$

$$= \frac{7 - 5}{10}$$

$$= \frac{2}{10}$$

$$= \frac{1}{5}$$

j. $m = \frac{7}{8}$ is not a solution to the equation $\frac{3}{4} + \frac{4}{5}m = \frac{2}{3}m$ as substitution of $\frac{7}{8}$ for m in the equation does not result in equal values for the expressions on the left and right sides of the equation:

$$LS = \frac{3}{4} + \frac{4}{5}m$$

$$= \frac{3}{4} + \frac{4}{5} \times \frac{7}{8}$$

$$= \frac{3}{4} + \frac{7}{10}$$

$$= \frac{15 + 14}{20}$$

$$= \frac{29}{20}$$

but

$$RS = \frac{2}{3}m$$

$$= \frac{2}{3} \times \frac{7}{8}$$

$$= \frac{7}{12}$$

k. $p = 1$ and $t = 2$ is a solution to the equation $2p - 3t = -4$ as substitution of 1 for p and 2 for t in the equation results in equal values for the expressions on the left and right sides of the equation:

$$
\begin{aligned}
\text{LS} &= 2p - 3t \\
&= 2 \times 1 - 3 \times 2 \\
&= 2 - 6 \\
&= -4
\end{aligned}
$$

and

$$
\text{RS} = -4
$$

l. $n = -1$ and $m = 1$ is not a solution to the equation $-n + 2m = -3n + m$ as substitution of -1 for n and 1 for m in the equation does not result in equal values for the expressions on the left and right sides of the equation:

$$
\begin{aligned}
\text{LS} &= -n + 2m \\
&= -(-1) + 2(1) \\
&= 1 + 2 \\
&= 3
\end{aligned}
$$

but

$$
\begin{aligned}
\text{RS} &= -3n + m \\
&= -3(-1) + 1 \\
&= 3 + 1 \\
&= 4
\end{aligned}
$$

m. $t = 0$ and $v = 3$ is a solution to the equation $t = -v + 3$ as substitution of 0 for t and 3 for v in the equation results in equal values for the expressions on the left and right sides of the equation:

$$
\begin{aligned}
\text{LS} &= t \\
&= 0
\end{aligned}
$$

and

$$
\begin{aligned}
\text{RS} &= -v + 3 \\
&= -3 + 3 \\
&= 0
\end{aligned}
$$

Exercise Set 10.2

1. The equation is concrete.
2. The equation is concrete.
3. The equation is abstract.
4. The equation is abstract.
5. The equation is concrete.

6. The equation is abstract.
7. The equation is abstract.
8. The equation is concrete.
9. The equation is concrete.
10. The equation is abstract.

Exercise Set 10.3

1. $1.24m = 19.7 \times 0.77$

2. $3.2v = 4.5 \times 2.7$

3. $\frac{0.813}{t} = 1.22$

4. $4 - 3x = -2$

5. $2.4 \times \frac{v^2}{1.4} = 47.8$

6. $\frac{1}{2}\left(31.8^2\right)x = 130\,000$

7. $12.9 \times \frac{1}{2^n} = 8.7$

8. $-(4 - 2x) = 7$

9. $\frac{22.4 - v_0}{2.5 - 1.7} = 8.1$

10. $\frac{1}{2}(1.13)\,t^2 + 2.7t = 12.8$

11. $\frac{3020 - E_1}{5.7 - 1.8} = 1800$

12. $\frac{1}{2}\left(-\frac{2}{3} + x\right) = 0$

13. $\frac{1}{2}(2.8)(b_1 + 7.1) = 155.7$

Exercise Set 10.4

1. The equation $3x + 1 = 2$ analyzes into the two expressions $3x + 1$ and 2.

   The expression $3x + 1$ analyzes into the two terms $3x$ and 1. The term $3x$ analyzes into the two factors 3 and x. The term 1 analyzes into the single factor 1.

   The expression 2 analyzes into the single term 2 which analyzes into the single factor 2.

2. The equation $x^2 - 6x + 8 = 0$ analyzes into the two expressions $x^2 - 6x + 8$ and 0.

   The expression $x^2 - 6x + 8$ analyzes into the three terms x^2, $6x$, and 8. The term x^2 analyzes into the single factor x^2 which is an exponent problem. The term $6x$ analyzes into the two factors 6 and x. The term 8 analyzes into the single factor 8.

   The term 0 analyzes into the single term 0 which analyzes into the single factor 0.

3. The equation $-2x - 3y = 1$ analyzes into the two expressions $-2x - 3y$ and 1.

   The expression $-2x - 3y$ analyzes into the two terms $-2x$ and $3y$. The term $-2x$ analyzes into the two factors -2 and x. The term $3y$ analyzes into the two factors 3 and y.

The expression 1 analyzes into the single term 1 which analyzes into the single factor 1.

4. The equation $x = -y + 2$ analyzes into the two expressions x and $-y + 2$.

The expression x analyzes into the single term x which analyzes into the single factor x.

The expression $-y + 2$ analyzes into the two terms $-y$ and 2. The term $-y$ analyzes into the two factors -1 and y. The term 2 analyzes into the single factor 2.

5. The equation $xy = 24$ analyzes into the two expressions xy and 24.

The expression xy analyzes into the single term xy which analyzes into the two factors x and y.

The expression 24 analyzes into the single term 24 which analyzes into the single factor 24.

6. The equation $x^2 + 2xy + y^2 = 0$ analyzes into the two expressions $x^2 + 2xy + y^2$ and 0.

The expression $x^2 + 2xy + y^2$ analyzes into the three terms x^2, $2xy$ and y^2. The term x^2 analyzes into the single factor x^2 which is an exponent problem. The term $2xy$ analyzes into the three factors 2, x, and y. The term y^2 analyzes into the single factor y^2 which is an exponent problem.

The expression 0 analyzes into the single term 0 which analyzes into the single factor 0.

7. The equation $z = y(y - 1)$ analyzes into the two expressions z and $y(y - 1)$.

The expression z analyzes into the single term z which analyzes into the single factor z.

The expression $y(y - 1)$ analyzes into the single term $y(y - 1)$ which analyzes into the two factors y and $y - 1$. The factor $y - 1$ analyzes into the two terms y and 1.

8. The equation $y = (x - 2)^2$ analyzes into the two expressions y and $(x - 2)^2$.

The term y analyzes into the single term y which analyzes into the single factor y.

The term $(x - 2)^2$ analyzes into the single term $(x - 2)^2$ which analyzes into the single factor $(x - 2)^2$ which is an exponent problem. The exponent problem analyzes into its base $x - 2$ and the exponent 2. The base analyzes into the two terms x and 2.

9. The equation $2x^2 + 3xy + 3y^2 - 6y = 0$ analyzes into the two expressions $2x^2 + 3xy + 3y^2 - 6y$ and 0.

The expression $2x^2 + 3xy + 3y^2 - 6y$ analyzes into the four terms $2x^2$, $3xy$, $3y^2$, and $6y$. The term $2x^2$ analyzes into the single term $2x^2$ which analyzes into the two factors 2 and x^2 the last of which is an exponent problem. The term $3xy$ analyzes into the three factors 3, x, and y. The term $3y^2$ analyzes into the two factors 3 and y^2 the last of which is an exponent problem. The term $6y$ analyzes into the two factors 6 and y.

The term 0 analyzes into the single term 0 which analyzes into the single factor 0.

10. The equation $-\frac{1}{2}x = \frac{3}{5}$ analyzes into the two expressions $-\frac{1}{2}x$ and $\frac{3}{5}$.

The expression $-\frac{1}{2}x$ analyzes into the single term $-\frac{1}{2}x$ which analyzes into the two factors $-\frac{1}{2}$ and x.

The expression $\frac{3}{5}$ analyzes into the single term $\frac{3}{5}$ which analyzes into the single factor $\frac{3}{5}$.

11. The equation $\frac{3}{4}x - \frac{2}{3}y = 0$ analyzes into the two expressions $\frac{3}{4}x - \frac{2}{3}y$ and 0.

The expression $\frac{3}{4}x - \frac{2}{3}y$ analyzes into the two terms $\frac{3}{4}x$ and $\frac{2}{3}y$. The term $\frac{3}{4}x$ analyzes into the two factors $\frac{3}{4}$ and x. The term $\frac{2}{3}y$ analyzes into the two factors $\frac{2}{3}$ and y.

The expression 0 analyzes into the single term 0 which analyzes into the single factor 0.

12. The equation $-(3x - 2x) - 5x = 2(3x + 2)$ analyzes into the two expressions $-(3x - 2x) - 5x$ and $2(3x + 2)$.

The expression $-(3x - 2x) - 5x$ analyzes into the two terms $-(3x - 2x)$ and $5x$. The term $-(3x - 2x)$ analyzes into the factors -1 and $3x - 2x$. The second factor analyzes into the two terms $3x$ which further analyzes into the two factors 3 and x, and $2x$ which further analyzes into the two factors 2 and x. The term $5x$ analyzes into the two factors 5 and x.

The expression $2(3x + 2)$ analyzes into the single term $2(3x + 2)$ which analyzes into the two factors 2 and $3x + 2$. The second factor analyzes into the two terms $3x$, which further analyzes into the two factors 3 and x, and 2 which analyzes into the single factor 2.

13. The equation $2(x - 7) = 1 - 5(x + 4)$ analyzes into the two expressions $2(x - 7)$ and $1 - 5(x + 4)$.

The expression $2(x - 7)$ analyzes into the single term $2(x - 7)$ which analyzes into the two factors 2 and $x - 7$ the second of which analyzes into the two terms

x and 7.

The expression $1 - 5\,(x + 4)$ analyzes into the two terms 1 which analyzes into the single factor 1, and $5\,(x + 4)$ which analyzes into the two factors 5 and $x + 4$ the second of which analyzes into the two terms x and 4.

14. The equation $\frac{1}{x} + \frac{1}{y} = \frac{1}{7}$ analyzes into the two expressions $\frac{1}{x} + \frac{1}{y}$ and $\frac{1}{7}$.

The expression $\frac{1}{x} + \frac{1}{y}$ analyzes into the two terms $\frac{1}{x}$ and $\frac{1}{y}$. The first term analyzes into the single factor $\frac{1}{x}$. This factor involves a division of 1 by x. The second term analyzes into the single factor $\frac{1}{y}$ which involves a division of 1 by y.

The expression $\frac{1}{7}$ analyzes into the single term $\frac{1}{7}$ which analyzes into the single factor $\frac{1}{7}$.

15. The equation $\frac{2}{1-x} - \frac{3}{2x+1} = 1$ analyzes into the two expressions $\frac{2}{1-x} - \frac{3}{2x+1}$ and 1.

The expression $\frac{2}{1-x} - \frac{3}{2x+1}$ analyzes into the two terms $\frac{2}{1-x}$ and $\frac{3}{2x+1}$. The first term analyzes into the single factor $\frac{2}{1-x}$ which involves a division of 2 by $1 - x$ which analyzes into the two terms 1 and x. The second term analyzes into the single factor $\frac{3}{2x+1}$ which involves a division of 3 and $2x + 1$ which analyzes into the two terms $2x$ which analyzes into the two factors 2 and x, and 1 which analyzes into the single factor 1.

The expression 1 analyzes into the single term 1 which analyzes into the single factor 1.

16. The equation $2\,(x + 1) - \log 32 = 3$ analyzes into the two expressions $2\,(x + 1) - \log 32$ and 3.

The expression $2\,(x + 1) - \log 32$ analyzes into the two terms $2\,(x + 1)$ which analyzes into the two factors 2 and $x + 1$ which further analyzes into the two terms x and 1, and $\log 32$ which analyzes into the single factor $\log 32$.

The expression 3 analyzes into the single term 3 which analyzes into the single factor 3.

17. The equation $2x + 3 \log 1 = -8$ analyzes into the two expressions $2x + 3 \log 1$ and -8.

The expression $2x + 3 \log 1$ analyzes into the two terms $2x$ which analyzes into the two factors 2 and x, and $3 \log 1$ which analyzes into the two factors 3 and $\log 1$.

The expression -8 analyzes into the single term -8 which analyzes into the single factor -8.

18. The equation $x - 2\sqrt{x} + 1 = 0$ analyzes into the two expressions $x - 2\sqrt{x} + 1$ and 0.

The expression $x - 2\sqrt{x} + 1$ analyzes into the three terms x, $2\sqrt{x}$ which analyzes into the two factors 2 and $\sqrt{x}$ the second if which is a root problem, and 1.

The expression 0 analyzes into the single term 0 which analyzes into the single factor 0.

19. The equation $2x\sqrt{x}-3 = 2\left(\sqrt{x}-1\right)$ analyzes into the two expressions $2x\sqrt{x}-3$ and $2\left(\sqrt{x}-1\right)$.

The expression $2x\sqrt{x}-3$ analyzes into the two terms $2x\sqrt{x}$ which analyzes into the three factors 2, x and $\sqrt{x}$ the last of which is a root problem, and 3 which analyzes into the single factor 3.

The expression $2\left(\sqrt{x}-1\right)$ analyzes into the single term $2\left(\sqrt{x}-1\right)$ which analyzes into the two factors 2 and $\sqrt{x}-1$ the second of which analyzes into the two terms $\sqrt{x}$ which analyzes into the single factor $\sqrt{x}$ which is a root problem, and 1 which analyzes into the single factor 1.

20. The equation $(x-y)^2+(x+y)^2 = 1$ analyzes into the two expressions $(x-y)^2+(x+y)^2$ and 1.

The expression $(x-y)^2+(x+y)^2$ analyzes into the two terms $(x-y)^2$ which is an exponent problem with base $x-y$ which analyzes into the two terms x and y and exponent 2, and $(x+y)^2$ which is an exponent problem with base $x+y$ which analyzes into the two terms x and y and exponent 2.

The expression 1 analyzes into the single term 1 which analyzes into the single factor 1.

Exercise Set 10.5

1. a.

$$2x + 3x$$
$$5x$$

g.

$$4x - 12x$$
$$-8x$$

b.

$$182x + 95x$$
$$277x$$

h.

$$118x - 210x$$
$$-92x$$

c.

$$15x + x$$
$$16x$$

i.

$$6x - x$$
$$5x$$

d.

$$y + 9y$$
$$10y$$

j.

$$z - 6z$$
$$-5z$$

e.

$$7z - 2z$$
$$5z$$

k.

$$-8x + 3x$$
$$-5x$$

f.

$$142x - 123x$$
$$19x$$

l.

$$-2x + 7x$$
$$5x$$

m.

$$-4y - 6y$$
$$-10y$$

n.

$$-5y - 2y$$
$$-7y$$

o.

$$\frac{2}{3}x + \frac{1}{2}x$$
$$\frac{4 + 3}{6}x$$
$$\frac{7}{6}x$$

p.

$$\frac{3}{5}x + \frac{5}{6}x$$
$$\frac{18 + 25}{30}x$$
$$\frac{43}{30}x$$

q.

$$\frac{4}{5}y + 2y$$
$$\frac{4}{5}y + \frac{2}{1}y$$
$$\frac{4 + 10}{5}x$$
$$\frac{14}{5}x$$

Alternative

$$\frac{4}{5}y + 2y$$
$$2\frac{4}{5}y$$
$$\frac{14}{5}x$$

r.

$$4x + \frac{2}{3}x$$
$$\frac{4}{1}x + \frac{2}{3}x$$
$$\frac{12 + 2}{3}x$$
$$\frac{14}{3}x$$

Alternative

$$4x + \frac{2}{3}x$$
$$4\frac{2}{3}x$$
$$\frac{14}{3}x$$

s.

$$\frac{-1}{4}y + \frac{2}{7}y$$
$$\frac{-7 + 8}{28}y$$
$$\frac{1}{28}y$$

t.

$$\frac{-2}{3}z - \frac{4}{5}z$$
$$\frac{-10 - 12}{15}z$$
$$\frac{-22}{15}z$$
$$-\frac{22}{15}z$$

u.

$$\frac{-4}{5}z + \frac{2}{3}z$$
$$\frac{-12 + 10}{15}z$$
$$\frac{-2}{15}z$$
$$-\frac{2}{15}z$$

v.

$$\frac{2}{7}x - \frac{3}{4}x$$
$$\frac{8 - 21}{28}x$$
$$\frac{-13}{28}x$$
$$-\frac{13}{28}x$$

w.

$$-\frac{4}{15}x - \frac{3}{10}x$$

$$\frac{-4}{15}x - \frac{3}{10}x$$

$$\frac{-8-9}{30}x$$

$$\frac{-17}{30}x$$

$$-\frac{17}{30}x$$

x.

$$-\frac{1}{6}x - \frac{3}{4}x$$

$$\frac{-1}{6}x - \frac{3}{4}x$$

$$\frac{-2-9}{12}x$$

$$\frac{-11}{12}x$$

$$-\frac{11}{12}x$$

2. a.

$$5x + 2x + 3x$$

$$10x$$

b.

$$217x + 82x + 318x$$

$$617x$$

c.

$$12x + 3x - 5x$$

$$10x$$

d.

$$y - 4y + 2y$$

$$- y$$

e.

$$- 16y - 7y - 32y + y$$

$$- 54y$$

f.

$$- 13z + 42z - 14z$$

$$15z$$

g.

$$\frac{5}{6}x + \frac{1}{3}x - \frac{1}{2}x$$

$$\frac{5+2-3}{6}x$$

$$\frac{4}{6}x$$

$$\frac{2}{3}x$$

h.

$$\frac{3}{4}x - \frac{1}{6}x - \frac{2}{3}x + \frac{1}{2}x$$

$$\frac{9-2-8+6}{12}x$$

$$\frac{5}{12}x$$

i.

$$\frac{-1}{3}y - \frac{3}{4}y - \frac{2}{5}y$$

$$\frac{-20-45-24}{60}y$$

$$\frac{-89}{60}y$$

$$-\frac{89}{60}y$$

j.

$$\frac{-4}{21}z - \frac{1}{14}z + \frac{5}{42}z$$

$$\frac{-8-3+5}{42}z$$

$$\frac{-6}{42}z$$

$$-\frac{1}{7}z$$

k.

$$\frac{1}{12}z + \frac{3}{8}z - \frac{5}{16}z$$

$$\frac{4+18-15}{48}z$$

$$\frac{7}{48}z$$

l.

$$-\frac{3}{8}y - \frac{1}{6}y - y + \frac{1}{2}y$$

$$\frac{-3}{8}y - \frac{1}{6}y - \frac{1}{1}y + \frac{1}{2}y$$

$$\frac{-9 - 4 - 24 + 12}{24}y$$

$$\frac{-25}{24}y$$

$$-\frac{25}{24}y$$

m.

$$\frac{1}{3}x + \frac{2}{3}x - \frac{1}{2}x + \frac{4}{5}x$$

$$\frac{10 + 20 - 15 + 24}{30}x$$

$$\frac{39}{30}x$$

$$\frac{13}{10}x$$

n.

$$\frac{3}{25}x + \frac{4}{15}x - \frac{3}{10}x$$

$$\frac{18 + 40 - 45}{150}x$$

$$\frac{13}{150}x$$

o.

$$\frac{2}{3}x - x + \frac{1}{2}x$$

$$\frac{2}{3}x - \frac{1}{1}x + \frac{1}{2}x$$

$$\frac{4 - 6 + 3}{6}x$$

$$\frac{1}{6}x$$

3. a.

$$-(-3)x + (-2)x$$

$$3x - 2x$$

$$x$$

b.

$$3y - (-4y)$$

$$3y + 4y$$

$$7y$$

c.

$$2z - (-3)z + 2z - 4z$$

$$2z + 3z + 2z - 4z$$

$$3z$$

d.

$$-(-2)x - 3x + (-3)x$$

$$2x - 3x - 3x$$

$$-4x$$

e.

$$-4x - (-3)x - (-5)x$$

$$-4x + 3x + 5x$$

$$4x$$

f.

$$-(-x) + (-3)x - (-2x)$$

$$x - 3x + 2x$$

$$0$$

g.

$$\frac{2}{3}x - \frac{-2}{5}x$$

$$\frac{2}{3}x + \frac{2}{5}x$$

$$\frac{10 + 6}{15}x$$

$$\frac{16}{15}x$$

h.

$$-\frac{-1}{2}y + \frac{-2}{-3}y$$

$$\frac{1}{2}y + \frac{2}{3}y$$

$$\frac{3 + 4}{6}y$$

$$\frac{7}{6}y$$

i.

$$- (-4)x - \frac{-1}{-5}x$$

$$4x - \frac{1}{5}x$$

$$\frac{4}{1}x - \frac{1}{5}x$$

$$\frac{20 - 1}{5}x$$

$$\frac{19}{5}x$$

Alternative

$$- (-4)x - \frac{-1}{-5}x$$

$$4x - \frac{1}{5}x$$

$$3\frac{4}{5}x$$

$$\frac{19}{5}x$$

j.

$$\frac{-2}{-21}x - \frac{3}{-14}x$$

$$\frac{2}{21}x + \frac{3}{14}x$$

$$\frac{4 + 9}{42}x$$

$$\frac{13}{42}x$$

k.

$$- (-1)z + \frac{3}{4}z + 3z$$

$$z + \frac{3}{4}z + 3z$$

$$\frac{1}{1}z + \frac{3}{4}z + \frac{3}{1}z$$

$$\frac{4 + 3 + 12}{4}z$$

$$\frac{19}{4}z$$

Alternative

$$- (-1)z + \frac{3}{4}z + 3z$$

$$z + \frac{3}{4}z + 3z$$

$$4\frac{3}{4}z$$

$$\frac{19}{4}z$$

l.

$$- \frac{-3}{5}y + \frac{2}{-3}y - \frac{-3}{-4}y$$

$$\frac{3}{5}y - \frac{2}{3}y - \frac{3}{4}y$$

$$\frac{36 - 40 - 45}{60}y$$

$$\frac{-49}{60}y$$

$$- \frac{49}{60}y$$

4. a.

$$2(-3)x + (-2)(-4)x$$

$$- 6x + 8x$$

$$2x$$

b.

$$- 3(4)y - (-5)y$$

$$- 12y + 5y$$

$$- 7y$$

c.

$$- 4z - (-2)(-1)z - 7z$$

$$- 4z - 2z - 7z$$

$$- 13z$$

d.

$$9x - 2(-1)(-3)x + (-1)x$$

$$9x - 6x - x$$

$$2x$$

e.

$$-4\,(-3)\,(2)\,y\,-\,y$$
$$24y\,-\,y$$
$$23y$$

f.

$$-2\,(-4x)\,+\,(-3)\,(-2)\,x$$
$$8x\,+\,6x$$
$$14x$$

g.

$$\frac{2}{3}x\,-\,\frac{-2}{5}x$$
$$\frac{2}{3}x\,+\,\frac{2}{5}x$$
$$\frac{10\,+\,6}{15}x$$
$$\frac{16}{15}x$$

h.

$$-\,\frac{-1}{2}y\,+\,\frac{-2}{-3}y$$
$$\frac{1}{2}y\,+\,\frac{2}{3}y$$
$$\frac{3\,+\,4}{6}y$$
$$\frac{7}{6}y$$

5. a.

$$4\times3x\,+\,2\times5x$$
$$12x\,+\,10x$$
$$22x$$

b.

$$-\,5\times2\times3x\,+\,x$$
$$-\,30x\,+\,x$$
$$-\,29x$$

c.

$$-\,8\times2x\,-\,5\times3x$$
$$-\,16x\,-\,15x$$
$$-\,31x$$

d.

$$4\,(3)\,x\,-\,2\,(3)\,x$$
$$12x\,-\,6x$$
$$6x$$

e.

$$-\,(-2)\,x\,+\,3\,(-2)\,(-1)\,x$$
$$2x\,+\,6x$$
$$8x$$

f.

$$2\,(-3)\,(-5)\,x\,+\,3\,(-4)\,x$$
$$30x\,-\,12x$$
$$18x$$

g.

$$\frac{1}{2}\times\frac{2}{3}x\,+\,\frac{3}{4}\times\frac{1}{3}x$$
$$\frac{1}{3}x\,+\,\frac{1}{4}x$$
$$\frac{4\,+\,3}{12}x$$
$$\frac{7}{12}x$$

h.

$$-\,\frac{3}{5}\times\frac{1}{2}x\,-\,\frac{1}{4}\times\frac{2}{3}x$$
$$-\,\frac{3}{10}x\,-\,\frac{1}{6}x$$
$$\frac{-3}{10}x\,-\,\frac{1}{6}x$$
$$\frac{-9\,-\,5}{30}x$$
$$\frac{-14}{30}x$$
$$-\,\frac{7}{15}x$$

i.

$$\frac{1}{2}\left(\frac{1}{3}\right)y - \frac{3}{5}\left(-\frac{1}{3}\right)y$$

$$\frac{1}{6}y + \frac{1}{5}y$$

$$\frac{5+6}{30}y$$

$$\frac{11}{30}y$$

j.

$$-\frac{3}{5}\left(\frac{-1}{3}\right)x - \left(-\frac{3}{4}\right)\left(\frac{1}{2}\right)x$$

$$\frac{1}{5}x + \frac{3}{8}x$$

$$\frac{8+15}{40}x$$

$$\frac{23}{40}x$$

6. a.

$$-3x + 2y - 4x + y$$

$$-7x + 3y$$

b.

$$-4 + 2x - 3$$

$$2x - 7$$

c.

$$2x - (-4)x + 3y - 2y$$

$$2x + 4x + 3y - 2y$$

$$6x + y$$

d.

$$-(-4y) + x - (-3)x - 2y$$

$$4y + x + 3x - 2y$$

$$4x + 2y$$

e.

$$\frac{1}{-2} + (-3)x - \frac{-2}{3}x - 4$$

$$\frac{-1}{2} - 3x + \frac{2}{3}x - 4$$

$$\frac{-1}{2} - \frac{3}{1}x + \frac{2}{3}x - \frac{4}{1}$$

$$\frac{-9+2}{3}x - \frac{1+8}{2}$$

$$\frac{-7}{3}x - \frac{9}{2}$$

$$-\frac{7}{3}x - \frac{9}{2}$$

Alternative

$$\frac{1}{-2} + (-3)x - \frac{-2}{3}x - 4$$

$$-\frac{1}{2} - 3x + \frac{2}{3}x - 4$$

$$-2\frac{1}{3}x - 4\frac{1}{2}$$

$$-\frac{7}{3}x - \frac{9}{2}$$

f.

$$-2x + (-3)y - 4x + (-y)$$

$$-2x - 3y - 4x - y$$

$$-6x - 4y$$

g.

$$\frac{-2}{3}y + \frac{-1}{-3} - \frac{1}{-2}y + \frac{4}{5}$$

$$\frac{-2}{3}y + \frac{1}{3} + \frac{1}{2}y + \frac{4}{5}$$

$$\frac{-4+3}{6}y + \frac{5+12}{15}$$

$$\frac{-1}{6}y + \frac{17}{15}$$

$$-\frac{1}{6}y + \frac{17}{15}$$

h.

$$2y - \frac{1}{-3} + (-3)\,y - \frac{1}{-2}y$$

$$2y + \frac{1}{3} - 3y + \frac{1}{2}y$$

$$\frac{2}{1}y + \frac{1}{3} - \frac{3}{1}y + \frac{1}{2}y$$

$$\frac{4 - 6 + 1}{2}y + \frac{1}{3}$$

$$\frac{-1}{2}y + \frac{1}{3}$$

$$-\frac{1}{2}y + \frac{1}{3}$$

Alternative

$$2y - \frac{1}{-3} + (-3)\,y - \frac{1}{-2}y$$

$$2y + \frac{1}{3} - 3y + \frac{1}{2}y$$

$$-\frac{1}{2}y + \frac{1}{3}$$

i.

$$\frac{1}{2}x + \frac{2}{3}y - 1 - x$$

$$\frac{1}{2}x + \frac{2}{3}y - 1 - \frac{1}{1}x$$

$$\frac{1 - 2}{2}x + \frac{2}{3}y - 1$$

$$\frac{-1}{2}x + \frac{2}{3}y - 1$$

$$-\frac{1}{2}x + \frac{2}{3}y - 1$$

Alternative

$$\frac{1}{2}x + \frac{2}{3}y - 1 - x$$

$$-\frac{1}{2}x + \frac{2}{3}y - 1$$

j.

$$-\frac{3}{-4}x + \frac{-1}{2}z - \frac{3}{-5}z + \frac{1}{3}x$$

$$\frac{3}{4}x - \frac{1}{2}z + \frac{3}{5}z + \frac{1}{3}x$$

$$\frac{9 + 4}{12}x + \frac{-5 + 6}{10}z$$

$$\frac{13}{12}x + \frac{1}{10}z$$

k.

$$\frac{-3}{4}x + \frac{1}{-4}y + \frac{-1}{-5}x + \frac{1}{-4}y$$

$$\frac{-3}{4}x - \frac{1}{4}y + \frac{1}{5}x - \frac{1}{4}y$$

$$\frac{-15 + 4}{20}x - \frac{1 + 1}{4}y$$

$$\frac{-11}{20}x - \frac{2}{4}y$$

$$-\frac{11}{20}x - \frac{1}{2}y$$

7. a.

$$2\,(3x + 4)$$

$$6x + 8$$

b.

$$-3\,(-2x - 5) + 3$$

$$6x + 15 + 3$$

$$6x + 18$$

c.

$$4\,(x + 2) - 3\,(2x - 4)$$

$$4x + 8 - 6x + 12$$

$$-2x + 20$$

d.

$$-2\,(3 - 4x) + 2x$$

$$-6 + 8x + 2x$$

$$10x - 6$$

e.

$$2y - 3\,(4y - 2) - 3$$

$$2y - 12y + 6 - 3$$

$$-10y + 3$$

f.

$$4z - 4\,(2x + 6)$$

$$4z - 8x - 24$$

$$-8x + 4z - 24$$

g.

$$-3x + 2(5x - 2)$$
$$-3x + 10x - 4$$
$$7x - 4$$

h.

$$\frac{1}{2}\left(x - \frac{2}{3}\right) - \frac{3}{4}\left(x + \frac{1}{9}\right)$$
$$\frac{1}{2}x - \frac{1}{3} - \frac{3}{4}x - \frac{1}{12}$$
$$\frac{2 - 3}{4}x - \frac{4 + 1}{12}$$
$$\frac{-1}{4}x - \frac{5}{12}$$
$$-\frac{1}{4}x - \frac{5}{12}$$

i.

$$-\frac{3}{5} + 3\left(\frac{1}{5} - \frac{2}{3}x\right)$$
$$-\frac{3}{5} + \frac{3}{1}\left(\frac{1}{5} - \frac{2}{3}x\right)$$
$$\frac{-3}{5} + \frac{3}{5} - \frac{2}{1}x$$
$$-2x + \frac{-3 + 3}{5}$$
$$-2x + \frac{0}{5}$$
$$-2x + 0$$
$$-2x$$

Alternative

$$-\frac{3}{5} + 3\left(\frac{1}{5} - \frac{2}{3}x\right)$$
$$-\frac{3}{5} + \frac{3}{5} - 2x$$
$$-2x$$

j.

$$\frac{1}{4}z + \frac{1}{5}\left(\frac{5}{6}z - \frac{2}{3}\right) - \frac{1}{4}$$
$$\frac{1}{4}z + \frac{1}{6}z - \frac{2}{15} - \frac{1}{4}$$
$$\frac{3 + 2}{12}z - \frac{8 + 15}{60}$$
$$\frac{5}{12}z - \frac{23}{60}$$

k.

$$-3(2x + 3y) - 3(5x - y)$$
$$-6x - 9y - 15x + 3y$$
$$-21x - 6y$$

l.

$$4(-x + y) + 5(x - y)$$
$$-4x + 4y + 5x - 5y$$
$$x - y$$

m.

$$\frac{1}{2}(x + 3y) - \frac{2}{3}(-2x + y)$$
$$\frac{1}{2}x + \frac{3}{2}y + \frac{4}{3}x - \frac{2}{3}y$$
$$\frac{3 + 8}{6}x + \frac{9 - 4}{6}y$$
$$\frac{11}{6}x + \frac{5}{6}y$$

n.

$$-\frac{3}{4}\left(\frac{1}{2}x + \frac{2}{3}y\right) - \frac{1}{3}\left(\frac{3}{5}x + \frac{1}{4}y\right)$$
$$-\frac{3}{8}x - \frac{1}{2}y - \frac{1}{5}x - \frac{1}{12}y$$
$$\frac{-3}{8}x - \frac{1}{2}y - \frac{1}{5}x - \frac{1}{12}y$$
$$\frac{-15 - 8}{40}x - \frac{6 + 1}{12}y$$
$$\frac{-23}{40}x - \frac{7}{12}y$$
$$-\frac{23}{40}x - \frac{7}{12}y$$

8. a.

$$3x + 2x - 1 = 5 + 2$$
$$5x - 1 = 7$$

b.

$$-5x + 3 - x = 4x - 3$$
$$-6x + 3 = 4x - 3$$

c.

$$-(-3)x + (-2)y = x + (-4)y$$
$$3x - 2y = x - 4y$$

d.

$$4x - (-2)y - (-3)x = y$$
$$4x + 2y + 3x = y$$
$$7x + 2y = y$$

e.

$$4(-3)x + (-2)(-3)x = 0$$
$$-12x + 6x = 0$$
$$-6x = 0$$

f.

$$\frac{2}{3}x + \frac{3}{4}x = \frac{-1}{2}x - \frac{2}{5}x$$
$$\frac{8+9}{12}x = \frac{-5-4}{10}x$$
$$\frac{17}{12}x = \frac{-9}{10}x$$
$$\frac{17}{12}x = -\frac{9}{10}x$$

g.

$$-\frac{3}{5}z + \frac{1}{2}z = \frac{3}{2}z - \frac{1}{4}z$$
$$\frac{-3}{5}z + \frac{1}{2}z = \frac{6-1}{4}z$$
$$\frac{-6+5}{10}z = \frac{5}{4}z$$
$$\frac{-1}{10}z = \frac{5}{4}z$$
$$-\frac{1}{10}z = \frac{5}{4}z$$

h.

$$-\frac{-1}{2}y - \frac{2}{3}y = \frac{-1}{-3}y + \frac{-3}{4}y$$

$$\frac{1}{2}y - \frac{2}{3}y = \frac{1}{3}y - \frac{3}{4}y$$

$$\frac{3-4}{6}y = \frac{4-9}{12}y$$

$$\frac{-1}{6}y = \frac{-5}{12}y$$

$$-\frac{1}{6}y = -\frac{5}{12}y$$

i.

$$3(x+1) - 2x = -3(x-2)$$

$$3x + 3 - 2x = -3x + 6$$

$$x + 3 = -3x + 6$$

j.

$$-(3x+1) + 4(1-x) = 0$$

$$-3x - 1 + 4 - 4x = 0$$

$$-7x + 3 = 0$$

k.

$$3(x+y) - 2(-x+y) = 0$$

$$3x + 3y + 2x - 2y = 0$$

$$5x + y = 0$$

l.

$$-2(3x-2y) - 3(x+2y) = 3(-x+5y)$$

$$-6x + 4y - 3x - 6y = -3x + 15y$$

$$-9x - 2y = -3x + 15y$$

m.

$$\frac{1}{3}(x-1) + \frac{2}{5}(x+1) = \frac{3}{4}(-x+1) - \frac{1}{4}(x-1)$$

$$\frac{1}{3}x - \frac{1}{3} + \frac{2}{5}x + \frac{2}{5} = -\frac{3}{4}x + \frac{3}{4} - \frac{1}{4}x + \frac{1}{4}$$

$$\frac{5+6}{15}x + \frac{-5+6}{15} = -\frac{3+1}{4}x + \frac{3+1}{4}$$

$$\frac{11}{15}x + \frac{1}{15} = -\frac{4}{4}x + \frac{4}{4}$$

$$\frac{11}{15}x + \frac{1}{15} = -x + 1$$

n.

$$\frac{1}{4}\left(\frac{1}{2}x - \frac{2}{3}\right) + \frac{2}{3}x = \frac{-1}{4}\left(-\frac{1}{3}x + \frac{2}{5}\right) - \frac{1}{2}$$

$$\frac{1}{8}x - \frac{1}{6} + \frac{2}{3}x = \frac{1}{12}x - \frac{1}{10} - \frac{1}{2}$$

$$\frac{3+16}{24}x - \frac{1}{6} = \frac{1}{12}x - \frac{1+5}{10}$$

$$\frac{19}{24}x - \frac{1}{6} = \frac{1}{12}x - \frac{6}{10}$$

$$\frac{19}{24}x - \frac{1}{6} = \frac{1}{12}x - \frac{3}{5}$$

Exercise Set 10.6.1

1.

$$x + 7 = 8$$
$$x = 8 - 7$$
$$x = 1$$

2.

$$x + 24 = 9$$
$$x = 9 - 24$$
$$x = -15$$

3.

$$x + \frac{1}{2} = \frac{3}{5}$$
$$x = \frac{3}{5} - \frac{1}{2}$$
$$x = \frac{6-5}{10}$$
$$x = \frac{1}{10}$$

4.

$$x + \frac{2}{3} = 3$$
$$x = 3 - \frac{2}{3}$$
$$x = \frac{3}{1} - \frac{2}{3}$$
$$x = \frac{9-2}{3}$$
$$x = \frac{7}{3}$$

Alternative

$$x + \frac{2}{3} = 3$$
$$x = 3 - \frac{2}{3}$$
$$x = 2\frac{1}{3}$$
$$x = \frac{7}{3}$$

5.

$$16 + x = 42$$
$$x = 42 - 16$$
$$x = 26$$

6.

$$2 + x = -1$$
$$x = -1 - 2$$
$$x = -3$$

7.

$$\frac{1}{4} + x = -\frac{1}{2}$$
$$x = -\frac{1}{2} - \frac{1}{4}$$
$$x = \frac{-1}{2} - \frac{1}{4}$$
$$x = \frac{-2-1}{4}$$
$$x = \frac{-3}{4}$$
$$x = -\frac{3}{4}$$

8.

$$\frac{2}{5} + x = \frac{3}{7}$$

$$x = \frac{3}{7} - \frac{2}{5}$$

$$x = \frac{15 - 14}{35}$$

$$x = \frac{1}{35}$$

9.

$$x - 3 = 7$$

$$x = 7 + 3$$

$$x = 10$$

10.

$$x - 18 = 5$$

$$x = 5 + 18$$

$$x = 23$$

11.

$$x - 1 = -6$$

$$x = -6 + 1$$

$$x = -5$$

12.

$$x - \frac{3}{5} = \frac{1}{4}$$

$$x = \frac{1}{4} + \frac{3}{5}$$

$$x = \frac{5 + 12}{20}$$

$$x = \frac{17}{20}$$

13.

$$-5 + x = 2$$

$$x = 2 + 5$$

$$x = 7$$

14.

$$-8 + x = -1$$

$$x = -1 + 8$$

$$x = 7$$

15.

$$-\frac{3}{5} + x = \frac{2}{7}$$

$$x = \frac{2}{7} + \frac{3}{5}$$

$$x = \frac{10 + 21}{35}$$

$$x = \frac{31}{35}$$

16.

$$\frac{-1}{4} + x = -\frac{1}{3}$$

$$x = -\frac{1}{3} + \frac{1}{4}$$

$$x = \frac{-1}{3} + \frac{1}{4}$$

$$x = \frac{-4 + 3}{12}$$

$$x = \frac{-1}{12}$$

$$x = -\frac{1}{12}$$

17.

$$4x = 28$$

$$x = \frac{28}{4}$$

$$x = 7$$

18.

$$3x = -6$$

$$x = \frac{-6}{3}$$

$$x = -2$$

19.

$$-5x = 12$$

$$x = \frac{12}{-5}$$

$$x = -\frac{12}{5}$$

20.

$$-2x = 22$$

$$x = \frac{22}{-2}$$

$$x = -11$$

21.
$$-7x = -15$$
$$x = \frac{-15}{-7}$$
$$x = \frac{15}{7}$$

22.
$$\frac{3}{4}x = \frac{1}{2}$$
$$x = \frac{4}{3} \times \frac{1}{2}$$
$$x = \frac{2}{3}$$

23.
$$-\frac{1}{3}x = \frac{2}{5}$$
$$x = -\frac{3}{1} \times \frac{2}{5}$$
$$x = -\frac{6}{5}$$

24.
$$\frac{4}{7}x = -\frac{1}{2}$$
$$x = \frac{7}{4}\left(-\frac{1}{2}\right)$$
$$x = -\frac{7}{8}$$

25.
$$\frac{-3}{8}x = -\frac{1}{4}$$
$$x = \frac{8}{-3}\left(-\frac{1}{4}\right)$$
$$x = \frac{2}{3}$$

26.
$$\frac{x}{4} = \frac{3}{5}$$
$$x = 4 \times \frac{3}{5}$$
$$x = \frac{4}{1} \times \frac{3}{5}$$
$$x = \frac{12}{5}$$

27.
$$\frac{x}{3} = 2$$
$$x = 3 \times 2$$
$$x = 6$$

28.
$$\frac{x}{-2} = \frac{3}{7}$$
$$x = -2 \times \frac{3}{7}$$
$$x = -\frac{2}{1} \times \frac{3}{7}$$
$$x = -\frac{6}{7}$$

29.
$$\frac{x}{-4} = -3$$
$$x = -4(-3)$$
$$x = 12$$

30.
$$\frac{x}{-8} = \frac{3}{5}$$
$$x = -8 \times \frac{3}{5}$$
$$x = -\frac{8}{1} \times \frac{3}{5}$$
$$x = -\frac{24}{5}$$

31.
$$x^2 = 25$$
$$x = \pm\sqrt{25}$$
$$x = \pm 5$$

32.
$$x^2 = 7$$
$$x = \pm\sqrt{7}$$

33.
$$x^2 = 0$$
$$x = \sqrt{0}$$
$$x = 0$$

34.

$$x^2 = -9$$

$$x = \pm\sqrt{-9}$$

x is undefined.[4]
No solutions.

35.

$$x^2 = \frac{1}{4}$$

$$x = \pm\sqrt{\frac{1}{4}}$$

$$x = \pm\frac{1}{2}$$

36.

$$x^2 = \frac{4}{9}$$

$$x = \pm\sqrt{\frac{4}{9}}$$

$$x = \pm\frac{2}{3}$$

37.

$$x^3 = 1$$

$$x = \sqrt[3]{1}$$

$$x = 1$$

38.

$$x^3 = -8$$

$$x = \sqrt[3]{-8}$$

$$x = -2$$

39.

$$x^3 = 2$$

$$x = \sqrt[3]{2}$$

40.

$$x^3 = -64$$

$$x = \sqrt[3]{-64}$$

$$x = -4$$

41.

$$x^3 = -\frac{8}{27}$$

$$x = \sqrt[3]{-\frac{8}{27}}$$

$$x = -\frac{2}{3}$$

42.

$$x^4 = 16$$

$$x = \pm\sqrt[4]{16}$$

$$x = \pm 2$$

43.

$$x^4 = -16$$

$$x = \pm\sqrt[4]{-16}$$

x is undefined.[5]
No solutions.

44.

$$x^5 = 32$$

$$x = \sqrt[5]{32}$$

$$x = 2$$

45.

$$x^5 = -32$$

$$x = \sqrt[5]{-32}$$

$$x = -2$$

46.

$$x^8 = 1$$

$$x = \pm\sqrt[8]{1}$$

$$x = \pm 1$$

47.

$$x^{100} = 1$$

$$x = \pm\sqrt[100]{1}$$

$$x = \pm 1$$

---

[4]Note that we are limiting ourselves to the set of real numbers, $\mathbb{R}$, in this book. This equation does have solutions in the set of complex numbers, $\mathbb{C}$.

[5]Note that we are limiting ourselves to the set of real numbers, $\mathbb{R}$, in this book. This equation does have solutions in the set of complex numbers, $\mathbb{C}$.

48.

$$x^{100} = -1$$
$$x = \pm \sqrt[100]{-1}$$

x is undefined.[6]
No solutions.

49.

$$x^{101} = 1$$
$$x = \sqrt[101]{1}$$
$$x = 1$$

50.

$$x^{101} = -1$$
$$x = \sqrt[101]{-1}$$
$$x = -1$$

51.

$$2^x = 8$$
$$x = \log_2 8$$
$$x = 3$$

52.

$$2^x = 1$$
$$x = \log_2 1$$
$$x = 0$$

53.

$$3^x = 81$$
$$x = \log_3 81$$
$$x = 4$$

54.

$$3^x = -3$$
$$x = \log_3 (-3)$$

x is undefined.
No solutions.

55.

$$3^x = \frac{1}{3}$$
$$x = \log_3 \frac{1}{3}$$
$$x = -1$$

56.

$$4^x = \frac{1}{16}$$
$$x = \log_4 \frac{1}{16}$$
$$x = -2$$

57.

$$10^x = 10\,000$$
$$x = \log 10\,000$$
$$x = 4$$

58.

$$2^x = 3$$
$$x = \log_2 3$$

59.

$$10^x = 4$$
$$x = \log 4$$

Exercise Set 10.6.2

1.

$$x + 7 = 8$$
$$x + 7 - 7 = 8 - 7$$
$$x = 1$$

2.

$$x + 24 = 9$$
$$x + 24 - 24 = 9 - 24$$
$$x = -15$$

---

[6]Note that we are limiting ourselves to the set of real numbers, $\mathbb{R}$, in this book. This equation does have solutions in the set of complex numbers, $\mathbb{C}$.

3.

$$x + \frac{1}{2} = \frac{3}{5}$$

$$x + \frac{1}{2} - \frac{1}{2} = \frac{3}{5} - \frac{1}{2}$$

$$x = \frac{6 - 5}{10}$$

$$x = \frac{1}{10}$$

4.

$$x + \frac{2}{3} = 3$$

$$x + \frac{2}{3} - \frac{2}{3} = 3 - \frac{2}{3}$$

$$x = \frac{3}{1} - \frac{2}{3}$$

$$x = \frac{9 - 2}{3}$$

$$x = \frac{7}{3}$$

Alternative

$$x + \frac{2}{3} = 3$$

$$x + \frac{2}{3} - \frac{2}{3} = 3 - \frac{2}{3}$$

$$x = 2\frac{1}{3}$$

$$x = \frac{7}{3}$$

5.

$$16 + x = 42$$

$$16 + x - 16 = 42 - 16$$

$$x = 26$$

6.

$$2 + x = -1$$

$$2 + x - 2 = -1 - 2$$

$$x = -3$$

7.

$$\frac{1}{4} + x = -\frac{1}{2}$$

$$\frac{1}{4} + x - \frac{1}{4} = -\frac{1}{2} - \frac{1}{4}$$

$$x = \frac{-1}{2} - \frac{1}{4}$$

$$x = \frac{-2 - 1}{4}$$

$$x = \frac{-3}{4}$$

$$x = -\frac{3}{4}$$

8.

$$\frac{2}{5} + x = \frac{3}{7}$$

$$\frac{2}{5} + x - \frac{2}{5} = \frac{3}{7} - \frac{2}{5}$$

$$x = \frac{15 - 14}{35}$$

$$x = \frac{1}{35}$$

9.

$$x - 3 = 7$$

$$x - 3 + 3 = 7 + 3$$

$$x = 10$$

10.

$$x - 18 = 5$$

$$x - 18 + 18 = 5 + 18$$

$$x = 23$$

11.

$$x - 1 = -6$$

$$x - 1 + 1 = -6 + 1$$

$$x = -5$$

12.

$$x - \frac{3}{5} = \frac{1}{4}$$

$$x - \frac{3}{5} + \frac{3}{5} = \frac{1}{4} + \frac{3}{5}$$

$$x = \frac{5 + 12}{20}$$

$$x = \frac{17}{20}$$

13.

$$-5 + x = 2$$
$$-5 + x + 5 = 2 + 5$$
$$x = 7$$

14.

$$-8 + x = -1$$
$$-8 + x + 8 = -1 + 8$$
$$x = 7$$

15.

$$-\frac{3}{5} + x = \frac{2}{7}$$
$$-\frac{3}{5} + x + \frac{3}{5} = \frac{2}{7} + \frac{3}{5}$$
$$x = \frac{10 + 21}{35}$$
$$x = \frac{31}{35}$$

16.

$$\frac{-1}{4} + x = -\frac{1}{3}$$
$$\frac{-1}{4} + x + \frac{1}{4} = -\frac{1}{3} + \frac{1}{4}$$
$$x = \frac{-1}{3} + \frac{1}{4}$$
$$x = \frac{-4 + 3}{12}$$
$$x = \frac{-1}{12}$$
$$x = -\frac{1}{12}$$

17.

$$4x = 28$$
$$\frac{4x}{4} = \frac{28}{4}$$
$$x = 7$$

18.

$$3x = -6$$
$$\frac{3x}{3} = \frac{-6}{3}$$
$$x = -2$$

19.

$$-5x = 12$$
$$\frac{-5x}{-5} = \frac{12}{-5}$$
$$x = -\frac{12}{5}$$

20.

$$-2x = 22$$
$$\frac{-2x}{-2} = \frac{22}{-2}$$
$$x = -11$$

21.

$$-7x = -15$$
$$\frac{-7x}{-7} = \frac{-15}{-7}$$
$$x = \frac{15}{7}$$

22.

$$\frac{3}{4}x = \frac{1}{2}$$
$$\frac{4}{3} \times \frac{3}{4}x = \frac{4}{3} \times \frac{1}{2}$$
$$x = \frac{2}{3}$$

23.

$$-\frac{1}{3}x = \frac{2}{5}$$
$$-\frac{3}{1}\left(-\frac{1}{3}\right)x = -\frac{3}{1}\left(\frac{2}{5}\right)$$
$$x = -\frac{6}{5}$$

24.

$$\frac{4}{7}x = -\frac{1}{2}$$
$$\frac{7}{4}\left(\frac{4}{7}\right)x = \frac{7}{4}\left(-\frac{1}{2}\right)$$
$$x = -\frac{7}{8}$$

25.

$$\frac{-3}{8}x = -\frac{1}{4}$$

$$\frac{8}{-3}\left(\frac{-3}{8}\right)x = \frac{8}{-3}\left(-\frac{1}{4}\right)$$

$$x = \frac{2}{3}$$

26.

$$\frac{x}{4} = \frac{3}{5}$$

$$4 \times \frac{x}{4} = 4 \times \frac{3}{5}$$

$$x = \frac{4}{1} \times \frac{3}{5}$$

$$x = \frac{12}{5}$$

27.

$$\frac{x}{3} = 2$$

$$3 \times \frac{x}{3} = 3 \times 2$$

$$x = 6$$

28.

$$\frac{x}{-2} = \frac{3}{7}$$

$$-2 \times \frac{x}{-2} = -2 \times \frac{3}{7}$$

$$x = -\frac{2}{1} \times \frac{3}{7}$$

$$x = -\frac{6}{7}$$

29.

$$\frac{x}{-4} = -3$$

$$-4\left(\frac{x}{-4}\right) = -4(-3)$$

$$x = 12$$

30.

$$\frac{x}{-8} = \frac{3}{5}$$

$$-8 \times \frac{x}{-8} = -8 \times \frac{3}{5}$$

$$x = -\frac{8}{1} \times \frac{3}{5}$$

$$x = -\frac{24}{5}$$

31.

$$x^2 = 25$$

$$\sqrt{x^2} = \sqrt{25}$$

$$x = \pm 5$$

32.

$$x^2 = 7$$

$$\sqrt{x^2} = \sqrt{7}$$

$$x = \pm\sqrt{7}$$

33.

$$x^2 = 0$$

$$\sqrt{x^2} = \sqrt{0}$$

$$x = 0$$

34.

$$x^2 = -9$$

$$\sqrt{x^2} = \sqrt{-9}$$

x is undefined.[7]
No solutions.

35.

$$x^2 = \frac{1}{4}$$

$$\sqrt{x^2} = \sqrt{\frac{1}{4}}$$

$$x = \pm\frac{1}{2}$$

[7]Note that we are limiting ourselves to the set of real numbers, $\mathbb{R}$, in this book. This equation does have solutions in the set of complex numbers, $\mathbb{C}$.

36.
$$x^2 = \frac{4}{9}$$
$$\sqrt{x^2} = \sqrt{\frac{4}{9}}$$
$$x = \pm\frac{2}{3}$$

37.
$$x^3 = 1$$
$$\sqrt[3]{x^3} = \sqrt[3]{1}$$
$$x = 1$$

38.
$$x^3 = -8$$
$$\sqrt[3]{x^3} = \sqrt[3]{-8}$$
$$x = -2$$

39.
$$x^3 = 2$$
$$\sqrt[3]{x^3} = \sqrt[3]{2}$$
$$x = \sqrt[3]{2}$$

40.
$$x^3 = -64$$
$$\sqrt[3]{x^3} = \sqrt[3]{-64}$$
$$x = -4$$

41.
$$x^3 = -\frac{8}{27}$$
$$\sqrt[3]{x^3} = \sqrt[3]{-\frac{8}{27}}$$
$$x = -\frac{2}{3}$$

42.
$$x^4 = 16$$
$$\sqrt[4]{x^4} = \sqrt[4]{16}$$
$$x = \pm 2$$

43.
$$x^4 = -16$$
$$\sqrt[4]{x^4} = \sqrt[4]{-16}$$
x is undefined.[8]
No solutions.

44.
$$x^5 = 32$$
$$\sqrt[5]{x^5} = \sqrt[5]{32}$$
$$x = 2$$

45.
$$x^5 = -32$$
$$\sqrt[5]{x^5} = \sqrt[5]{-32}$$
$$x = -2$$

46.
$$x^8 = 1$$
$$\sqrt[8]{x^8} = \sqrt[8]{1}$$
$$x = \pm 1$$

47.
$$x^{100} = 1$$
$$\sqrt[100]{x^{100}} = \sqrt[100]{1}$$
$$x = \pm 1$$

48.
$$x^{100} = -1$$
$$\sqrt[100]{x^{100}} = \sqrt[100]{-1}$$
x is undefined.[9]
No solutions.

49.
$$x^{101} = 1$$
$$\sqrt[101]{x^{101}} = \sqrt[101]{1}$$
$$x = 1$$

---

[8]Note that we are limiting ourselves to the set of real numbers, $\mathbb{R}$, in this book. This equation does have solutions in the set of complex numbers, $\mathbb{C}$.

[9]Note that we are limiting ourselves to the set of real numbers, $\mathbb{R}$, in this book. This equation does have solutions in the set of complex numbers, $\mathbb{C}$.

50.

$$x^{101} = -1$$
$$\sqrt[101]{x^{101}} = \sqrt[101]{-1}$$
$$x = -1$$

51.

$$2^x = 8$$
$$\log_2 2^x = \log_2 8$$
$$x = 3$$

52.

$$2^x = 1$$
$$\log_2 2^x = \log_2 1$$
$$x = 0$$

53.

$$3^x = 81$$
$$\log_3 3^x = \log_3 81$$
$$x = 4$$

54.

$$3^x = -3$$
$$\log_3 3^x = \log_3 (-3)$$

x is undefined.
No solutions.

55.

$$3^x = \frac{1}{3}$$
$$\log_3 3^x = \log_3 \frac{1}{3}$$
$$x = -1$$

56.

$$4^x = \frac{1}{16}$$
$$\log_4 4^x = \log_4 \frac{1}{16}$$
$$x = -2$$

57.

$$10^x = 10\,000$$
$$\log 10^x = \log 10\,000$$
$$x = 4$$

58.

$$2^x = 3$$
$$\log_2 2^x = \log_2 3$$
$$x = \log_2 3$$

59.

$$10^x = 4$$
$$\log 10^x = \log 4$$
$$x = \log 4$$

Exercise Set 10.7

1.

$$x + 0.2 = 15.8$$
$$x = 15.8 - 0.2$$
$$x = 15.6$$

2.

$$x + 1.75 = 2.99$$
$$x = 2.99 - 1.75$$
$$x = 1.24$$

3.

$$18.2 + x = -16.7$$
$$x = -16.7 - 18.2$$
$$x = -34.9$$

4.

$$x - 3.7 = 5.6$$
$$x = 5.6 + 3.7$$
$$x = 9.3$$

5.

$$x - 12 = 17.9$$
$$x = 17.9 + 12$$
$$x = 29.9$$

6.

$$x - 8.8 = -6.4$$
$$x = -6.4 + 8.8$$
$$x = 2.4$$

7.

$$-6.4 + x = 9.6$$
$$x = 9.6 + 6.4$$
$$x = 16$$

8.

$$-8.76 + x = -1.97$$
$$x = -1.97 + 8.76$$
$$x = 6.79$$

9.

$$0.7x = 12$$
$$x = \frac{12}{0.7}$$
$$x = 17.142\,857\ldots$$

10.

$$7.9x = -6.2$$
$$x = \frac{-6.2}{7.9}$$
$$x = -0.784\,810\ldots$$

11.

$$-8.7x = -4.51$$
$$x = \frac{-4.51}{-8.7}$$
$$x = 0.518\,390\ldots$$

12.

$$\frac{x}{0.65} = 4.7$$
$$x = 0.65 \times 4.7$$
$$x = 3.055$$

13.

$$\frac{x}{-4.8} = 2.1$$
$$x = -4.8 \times 2.1$$
$$x = -10.08$$

14.

$$\frac{x}{-8.1} = -7.5$$
$$x = -8.1\,(-7.5)$$
$$x = 60.75$$

15.

$$x^2 = 0.25$$
$$x = \pm\sqrt{0.25}$$
$$x = \pm 0.5$$

16.

$$x^2 = 0.4$$
$$x = \pm\sqrt{0.4}$$
$$x = \pm 0.632\,455\ldots$$

17.

$$x^2 = 18.6$$
$$x = \pm\sqrt{18.6}$$
$$x = \pm 4.312\,771\ldots$$

18.

$$x^2 = -9.1$$
$$x = \pm\sqrt{-9.1}$$

x is undefined.[10]
No solutions.

19.

$$x^3 = 0.82$$
$$x = \sqrt[3]{0.82}$$
$$x = 0.935\,990\ldots$$

20.

$$x^3 = -7.7$$
$$x = \sqrt[3]{-7.7}$$
$$x = -1.974\,680\ldots$$

---

[10]Note that x is undefined in $\mathbb{R}$. This equation does have a solution in the set of complex numbers, $\mathbb{C}$.

21.

$$x^3 = 6.1$$
$$x = \sqrt[3]{6.1}$$
$$x = 1.827\,160\ldots$$

22.

$$x^3 = -6.1$$
$$x = \sqrt[3]{-6.1}$$
$$x = -1.827\,160\ldots$$

23.

$$x^4 = 0.0001$$
$$x = \pm\sqrt[4]{0.0001}$$
$$x = \pm0.1$$

24.

$$x^4 = 12.8$$
$$x = \pm\sqrt[4]{12.8}$$
$$x = \pm1.891\,483\ldots$$

25.

$$x^5 = -99.7$$
$$x = \sqrt[5]{-99.7}$$
$$x = -2.510\,377\ldots$$

26.

$$x^8 = -42.7$$
$$x = \pm\sqrt[8]{-42.7}$$

x is undefined.[11]
No solutions.

27.

$$x^{2.3} = 45.8$$
$$x = \sqrt[2.3]{45.8}$$
$$x = 5.273\,700\ldots$$

28.

$$x^{3.8} = 423$$
$$x = \sqrt[3.8]{423}$$
$$x = 4.910\,686\ldots$$

29.

$$10^x = 42.8$$
$$x = \log 42.8$$
$$x = 1.631\,443\ldots$$

30.

$$10^x = -15.8$$
$$x = \log -15.8$$
x is undefined.
No solutions.

31.

$$2^x = 0.68$$
$$x = \log_2 0.68$$
$$x = -0.556\,393\ldots$$

32.

$$2^x = 3.4$$
$$x = \log_2 3.4$$
$$x = 1.765\,534\ldots$$

33.

$$3^x = 72.6$$
$$x = \log_3 72.6$$
$$x = 3.900\,343\ldots$$

34.

$$3^x = -82.9$$
$$x = \log_3 -82.9$$
x is undefined.
No solutions.

35.

$$7.2^x = 182$$
$$x = \log_{7.2} 182$$
$$x = 2.636\,166\ldots$$

36.

$$2.6^x = 0.81$$
$$x = \log_{2.6} 0.81$$
$$x = -0.220\,532\ldots$$

---

[11]Note that x is undefined in $\mathbb{R}$. This equation does have a solution in the set of complex numbers, $\mathbb{C}$.

Exercise Set 11.2A

1.

$$-x = 14$$
$$x = -14$$

2.

$$-x = 5$$
$$x = -5$$

3.

$$-x = -2$$
$$x = 2$$

4.

$$-x = -7$$
$$x = 7$$

5.

$$-x = 0$$
$$x = 0$$

6.

$$4x = 24$$
$$x = \frac{24}{4}$$
$$x = 6$$

7.

$$-5x = 20$$
$$x = \frac{20}{-5}$$
$$x = -4$$

8.

$$-2x = 18$$
$$x = \frac{18}{-2}$$
$$x = -9$$

9.

$$-3x = -12$$
$$x = \frac{-12}{-3}$$
$$x = 4$$

10.

$$-9x = -18$$
$$x = \frac{-18}{-9}$$
$$x = 2$$

11.

$$3x = 2$$
$$x = \frac{2}{3}$$

12.

$$8x = 1$$
$$x = \frac{1}{8}$$

13.

$$3x = -2$$
$$x = \frac{-2}{3}$$
$$x = -\frac{2}{3}$$

14.

$$6x = -1$$
$$x = \frac{-1}{6}$$
$$x = -\frac{1}{6}$$

15.

$$-3x = 8$$
$$x = \frac{8}{-3}$$
$$x = -\frac{8}{3}$$

16.
$$-7x = 2$$
$$x = \frac{2}{-7}$$
$$x = -\frac{2}{7}$$

17.
$$-5x = -4$$
$$x = \frac{-4}{-5}$$
$$x = \frac{4}{5}$$

18.
$$-3x = -3$$
$$x = \frac{-3}{-3}$$
$$x = 1$$

19.
$$-4x = 0$$
$$x = \frac{0}{-4}$$
$$x = 0$$

20.
$$x + 2 = 18$$
$$x = 18 - 2$$
$$x = 16$$

21.
$$5 + x = -3$$
$$x = -3 - 5$$
$$x = -8$$

22.
$$x + 1 = 0$$
$$x = 0 - 1$$
$$x = -1$$

23.
$$x - 4 = 2$$
$$x = 2 + 4$$
$$x = 6$$

24.
$$3 - x = 7$$
$$-x = 7 - 3$$
$$-x = 4$$
$$x = -4$$

25.
$$-2 + 4x = 6$$
$$4x = 6 + 2$$
$$4x = 8$$
$$x = \frac{8}{4}$$
$$x = 2$$

26.
$$3x - 5 = 1$$
$$3x = 1 + 5$$
$$3x = 6$$
$$x = \frac{6}{3}$$
$$x = 2$$

27.
$$-2x - 6 = -14$$
$$-2x = -14 + 6$$
$$-2x = -8$$
$$x = \frac{-8}{-2}$$
$$x = 4$$

28.
$$4 = 3 + 2x$$
$$3 + 2x = 4$$
$$2x = 4 - 3$$
$$2x = 1$$
$$x = \frac{1}{2}$$

29.

$$5 = -x + 3$$
$$-x + 3 = 5$$
$$-x = 5 - 3$$
$$-x = 2$$
$$x = -2$$

30.

$$-7 + 8x = 0$$
$$8x = 0 + 7$$
$$8x = 7$$
$$x = \frac{7}{8}$$

31.

$$9 - 2x = 2$$
$$-2x = 2 - 9$$
$$-2x = -7$$
$$x = \frac{-7}{-2}$$
$$x = \frac{7}{2}$$

32.

$$6x + 1 = -3$$
$$6x = -3 - 1$$
$$6x = -4$$
$$x = \frac{-4}{6}$$
$$x = -\frac{2}{3}$$

33.

$$-8x - 6 = -7$$
$$-8x = -7 + 6$$
$$-8x = -1$$
$$x = \frac{-1}{-8}$$
$$x = \frac{1}{8}$$

Exercise Set 11.2B

1.

$$\frac{2}{3}x = 24$$
$$x = \frac{3}{2} \times 24$$
$$x = 36$$

2.

$$\frac{1}{3}x = 2$$
$$x = 3 \times 2$$
$$x = 6$$

3.

$$\frac{3}{5}x = -1$$
$$x = \frac{5}{3}(-1)$$
$$x = -\frac{5}{3}$$

4.

$$\frac{3}{7}x = 21$$
$$x = \frac{7}{3} \times 21$$
$$x = 49$$

5.

$$\frac{4}{5}x = 0$$
$$x = \frac{5}{4} \times 0$$
$$x = 0$$

6.

$$-\frac{2}{5}x = 8$$
$$x = -\frac{5}{2} \times 8$$
$$x = -20$$

7.

$$-\frac{3}{8}x = -5$$
$$x = -\frac{8}{3}(-5)$$
$$x = \frac{40}{3}$$

8.

$$\frac{5}{6}x = \frac{1}{4}$$
$$x = \frac{6}{5} \times \frac{1}{4}$$
$$x = \frac{3}{10}$$

9.

$$\frac{2}{3}x = \frac{6}{5}$$
$$x = \frac{3}{2} \times \frac{6}{5}$$
$$x = \frac{9}{5}$$

10.

$$\frac{-3}{4}x = \frac{-9}{8}$$
$$x = \frac{4}{-3} \times \frac{-9}{8}$$
$$x = \frac{3}{2}$$

11.

$$\frac{2}{-5}x = -\frac{1}{5}$$
$$x = \frac{-5}{2}\left(-\frac{1}{5}\right)$$
$$x = \frac{1}{2}$$

12.

$$3x = \frac{1}{2}$$
$$x = \frac{1}{3} \times \frac{1}{2}$$
$$x = \frac{1}{6}$$

13.

$$-2x = \frac{4}{5}$$
$$x = -\frac{1}{2} \times \frac{4}{5}$$
$$x = -\frac{2}{5}$$

14.

$$-x = \frac{5}{6}$$
$$x = -\frac{5}{6}$$

15.

$$9x = -\frac{3}{4}$$
$$x = \frac{1}{9}\left(-\frac{3}{4}\right)$$
$$x = -\frac{1}{12}$$

16.

$$\frac{3}{5}x + \frac{1}{2} = \frac{1}{3}$$
$$\frac{3}{5}x = \frac{1}{3} - \frac{1}{2}$$
$$\frac{3}{5}x = \frac{2 - 3}{6}$$
$$\frac{3}{5}x = \frac{-1}{6}$$
$$x = \frac{5}{3} \times \frac{-1}{6}$$
$$x = -\frac{5}{18}$$

17.

$$\frac{4}{7}x + \frac{3}{4} = \frac{2}{5}$$
$$\frac{4}{7}x = \frac{2}{5} - \frac{3}{4}$$
$$\frac{4}{7}x = \frac{8 - 15}{20}$$
$$\frac{4}{7}x = \frac{-7}{20}$$
$$x = \frac{7}{4} \times \frac{-7}{20}$$
$$x = -\frac{49}{80}$$

18.

$$-\frac{1}{3}x - \frac{2}{9} = -\frac{5}{6}$$
$$-\frac{1}{3}x = -\frac{5}{6} + \frac{2}{9}$$
$$-\frac{1}{3}x = \frac{-5}{6} + \frac{2}{9}$$
$$-\frac{1}{3}x = \frac{-15 + 4}{18}$$
$$-\frac{1}{3}x = \frac{-11}{18}$$
$$x = -\frac{3}{1} \times \frac{-11}{18}$$
$$x = \frac{11}{6}$$

19.

$$\frac{3}{5} + \frac{1}{8}x = \frac{1}{2}$$
$$\frac{1}{8}x = \frac{1}{2} - \frac{3}{5}$$
$$\frac{1}{8}x = \frac{5 - 6}{10}$$
$$\frac{1}{8}x = \frac{-1}{10}$$
$$x = \frac{8}{1} \times \frac{-1}{10}$$
$$x = -\frac{4}{5}$$

20.

$$-\frac{4}{9} + 2x = \frac{1}{3}$$
$$2x = \frac{1}{3} + \frac{4}{9}$$
$$2x = \frac{3 + 4}{9}$$
$$2x = \frac{7}{9}$$
$$x = \frac{1}{2} \times \frac{7}{9}$$
$$x = \frac{7}{18}$$

21.

$$-4x - \frac{2}{5} = \frac{1}{4}$$
$$-4x = \frac{1}{4} + \frac{2}{5}$$
$$-4x = \frac{5 + 8}{20}$$
$$-4x = \frac{13}{20}$$
$$x = -\frac{1}{4} \times \frac{13}{20}$$
$$x = -\frac{13}{80}$$

22.

$$-x + 1 = \frac{3}{4}$$
$$-x = \frac{3}{4} - 1$$
$$-x = \frac{3}{4} - \frac{1}{1}$$
$$-x = \frac{3 - 4}{4}$$
$$-x = \frac{-1}{4}$$
$$x = \frac{1}{4}$$

23.

$$2x - 3 = \frac{4}{5}$$
$$2x = \frac{4}{5} + 3$$
$$2x = \frac{4}{5} + \frac{3}{1}$$
$$2x = \frac{4 + 15}{5}$$
$$2x = \frac{19}{5}$$
$$x = \frac{1}{2} \times \frac{19}{5}$$
$$x = \frac{19}{10}$$

24.

$$3 - 5x = \frac{9}{2}$$
$$-5x = \frac{9}{2} - 3$$
$$-5x = \frac{9}{2} - \frac{3}{1}$$
$$-5x = \frac{9 - 6}{2}$$
$$-5x = \frac{3}{2}$$
$$x = -\frac{1}{5} \times \frac{3}{2}$$
$$x = -\frac{3}{10}$$

25.

$$-2 + \frac{2}{3}x = \frac{1}{4}$$
$$\frac{2}{3}x = \frac{1}{4} + 2$$
$$\frac{2}{3}x = \frac{1}{4} + \frac{2}{1}$$
$$\frac{2}{3}x = \frac{1 + 8}{4}$$
$$\frac{2}{3}x = \frac{9}{4}$$
$$x = \frac{3}{2} \times \frac{9}{4}$$
$$x = \frac{27}{8}$$

26.

$$-\frac{5}{6}x - \frac{3}{4} = 1$$
$$-\frac{5}{6}x = 1 + \frac{3}{4}$$
$$-\frac{5}{6}x = \frac{1}{1} + \frac{3}{4}$$
$$-\frac{5}{6}x = \frac{4 + 3}{4}$$
$$-\frac{5}{6}x = \frac{7}{4}$$
$$x = -\frac{6}{5} \times \frac{7}{4}$$
$$x = -\frac{21}{10}$$

27.

$$\frac{8}{3}x - \frac{8}{3} = 2$$
$$\frac{8}{3}x = 2 + \frac{8}{3}$$
$$\frac{8}{3}x = \frac{2}{1} + \frac{8}{3}$$
$$\frac{8}{3}x = \frac{6 + 8}{3}$$
$$\frac{8}{3}x = \frac{14}{3}$$
$$x = \frac{3}{8} \times \frac{14}{3}$$
$$x = \frac{7}{4}$$

28.

$$\frac{1}{3} = -2 + \frac{4}{7}x$$
$$-2 + \frac{4}{7}x = \frac{1}{3}$$
$$\frac{4}{7}x = \frac{1}{3} + 2$$
$$\frac{4}{7}x = \frac{1}{3} + \frac{2}{1}$$
$$\frac{4}{7}x = \frac{1 + 6}{3}$$
$$\frac{4}{7}x = \frac{7}{3}$$
$$x = \frac{7}{4} \times \frac{7}{3}$$
$$x = \frac{49}{12}$$

29.

$$-3 = -\frac{7}{5}x + 2$$
$$-\frac{7}{5}x + 2 = -3$$
$$-\frac{7}{5}x = -3 - 2$$
$$-\frac{7}{5}x = -5$$
$$x = -\frac{5}{7}(-5)$$
$$x = \frac{25}{7}$$

30.

$$-\frac{7}{8} = \frac{3}{5}x - \frac{1}{4}$$

$$\frac{3}{5}x - \frac{1}{4} = -\frac{7}{8}$$

$$\frac{3}{5}x = -\frac{7}{8} + \frac{1}{4}$$

$$\frac{3}{5}x = \frac{-7}{8} + \frac{1}{4}$$

$$\frac{3}{5}x = \frac{-7 + 2}{8}$$

$$\frac{3}{5}x = \frac{-5}{8}$$

$$x = \frac{5}{3} \times \frac{-5}{8}$$

$$x = -\frac{25}{24}$$

Exercise Set 11.2C

1.

$$4(x + 3) = 20$$

$$x + 3 = \frac{20}{4}$$

$$x + 3 = 5$$

$$x = 5 - 3$$

$$x = 2$$

2.

$$3(x + 5) = 24$$

$$x + 5 = \frac{24}{3}$$

$$x + 5 = 8$$

$$x = 8 - 5$$

$$x = 3$$

3.

$$2(8 + x) = 18$$

$$8 + x = \frac{18}{2}$$

$$8 + x = 9$$

$$x = 9 - 8$$

$$x = 1$$

4.

$$3(2 + x) = 6$$

$$2 + x = \frac{6}{3}$$

$$2 + x = 2$$

$$x = 2 - 2$$

$$x = 0$$

5.

$$5(x - 4) = 10$$

$$x - 4 = \frac{10}{5}$$

$$x - 4 = 2$$

$$x = 2 + 4$$

$$x = 6$$

6.

$$7(x - 2) = -7$$

$$x - 2 = \frac{-7}{7}$$

$$x - 2 = -1$$

$$x = -1 + 2$$

$$x = 1$$

7.

$$32 = 8\,(5 - y)$$
$$8\,(5 - y) = 32$$
$$5 - y = \frac{32}{8}$$
$$5 - y = 4$$
$$-y = 4 - 5$$
$$-y = -1$$
$$y = 1$$

8.

$$18 = 3\,(2 - y)$$
$$3\,(2 - y) = 18$$
$$2 - y = \frac{18}{3}$$
$$2 - y = 6$$
$$-y = 6 - 2$$
$$-y = 4$$
$$y = -4$$

9.

$$-2\,(y - 5) = 8$$
$$y - 5 = \frac{8}{-2}$$
$$y - 5 = -4$$
$$y = -4 + 5$$
$$y = 1$$

10.

$$-3\,(4 + y) = -21$$
$$4 + y = \frac{-21}{-3}$$
$$4 + y = 7$$
$$y = 7 - 4$$
$$y = 3$$

11.

$$-2\,(5 - y) = -6$$
$$5 - y = \frac{-6}{-2}$$
$$5 - y = 3$$
$$-y = 3 - 5$$
$$-y = -2$$
$$y = 2$$

12.

$$-(2y + 3) = -3$$
$$2y + 3 = 3$$
$$2y = 3 - 3$$
$$2y = 0$$
$$y = \frac{0}{2}$$
$$y = 0$$

13.

$$-(4 - 5z) = 6$$
$$4 - 5z = -6$$
$$-5z = -6 - 4$$
$$-5z = -10$$
$$z = \frac{-10}{-5}$$
$$z = 2$$

14.

$$5\,(3z - 1) = -5$$
$$3z - 1 = \frac{-5}{5}$$
$$3z - 1 = -1$$
$$3z = -1 + 1$$
$$3z = 0$$
$$z = \frac{0}{3}$$
$$z = 0$$

15.

$$-2\left(5 - 3z\right) = 8$$
$$5 - 3z = \frac{8}{-2}$$
$$5 - 3z = -4$$
$$-3z = -4 - 5$$
$$-3z = -9$$
$$z = \frac{-9}{-3}$$
$$z = 3$$

16.

$$3\left(4z + 2\right) = 7$$
$$4z + 2 = \frac{7}{3}$$
$$4z = \frac{7}{3} - 2$$
$$4z = \frac{7}{3} - \frac{2}{1}$$
$$4z = \frac{7 - 6}{3}$$
$$4z = \frac{1}{3}$$
$$z = \frac{1}{4} \times \frac{1}{3}$$
$$z = \frac{1}{12}$$

17.

$$-2\left(-z + 1\right) = 5$$
$$-z + 1 = \frac{5}{-2}$$
$$-z = \frac{5}{-2} - 1$$
$$-z = \frac{-5}{2} - \frac{1}{1}$$
$$-z = \frac{-5 - 2}{2}$$
$$-z = \frac{-7}{2}$$
$$z = \frac{7}{2}$$

18.

$$\frac{1}{2}\left(2x + 4\right) = 3$$
$$2x + 4 = 2 \times 3$$
$$2x + 4 = 6$$
$$2x = 6 - 4$$
$$2x = 2$$
$$x = \frac{2}{2}$$
$$x = 1$$

19.

$$\frac{2}{3}\left(\frac{3}{4}x - \frac{1}{2}\right) = \frac{1}{4}$$
$$\frac{3}{4}x - \frac{1}{2} = \frac{3}{2} \times \frac{1}{4}$$
$$\frac{3}{4}x - \frac{1}{2} = \frac{3}{8}$$
$$\frac{3}{4}x = \frac{3}{8} + \frac{1}{2}$$
$$\frac{3}{4}x = \frac{3 + 4}{8}$$
$$\frac{3}{4}x = \frac{7}{8}$$
$$x = \frac{4}{3} \times \frac{7}{8}$$
$$x = \frac{7}{6}$$

20.

$$\frac{3}{4}\left(\frac{5}{3}y - 1\right) = \frac{1}{3}$$

$$\frac{5}{3}y - 1 = \frac{4}{3} \times \frac{1}{3}$$

$$\frac{5}{3}y - 1 = \frac{4}{9}$$

$$\frac{5}{3}y = \frac{4}{9} + 1$$

$$\frac{5}{3}y = \frac{4}{9} + \frac{1}{1}$$

$$\frac{5}{3}y = \frac{4 + 9}{9}$$

$$\frac{5}{3}y = \frac{13}{9}$$

$$y = \frac{3}{5} \times \frac{13}{9}$$

$$x = \frac{13}{15}$$

21.

$$-\frac{2}{3}\left(z - \frac{3}{4}\right) = \frac{1}{5}$$

$$z - \frac{3}{4} = -\frac{3}{2} \times \frac{1}{5}$$

$$z - \frac{3}{4} = -\frac{3}{10}$$

$$z = -\frac{3}{10} + \frac{3}{4}$$

$$z = \frac{-3}{10} + \frac{3}{4}$$

$$z = \frac{-6 + 15}{20}$$

$$z = \frac{9}{20}$$

22.

$$4(x + 2) - 1 = 19$$

$$4(x + 2) = 19 + 1$$

$$4(x + 2) = 20$$

$$x + 2 = \frac{20}{4}$$

$$x + 2 = 5$$

$$x = 5 - 2$$

$$x = 3$$

23.

$$3(2x + 1) + 4 = 1$$

$$3(2x + 1) = 1 - 4$$

$$3(2x + 1) = -3$$

$$2x + 1 = \frac{-3}{3}$$

$$2x + 1 = -1$$

$$2x = -1 - 1$$

$$2x = -2$$

$$x = \frac{-2}{2}$$

$$x = -1$$

24.

$$-(3x - 1) = 0$$

$$3x - 1 = 0$$

$$3x = 0 + 1$$

$$3x = 1$$

$$x = \frac{1}{3}$$

25.

$$5\left[2(x - 1) + 3\right] = 45$$

$$2(x - 1) + 3 = \frac{45}{5}$$

$$2(x - 1) + 3 = 9$$

$$2(x - 1) = 9 - 3$$

$$2(x - 1) = 6$$

$$x - 1 = \frac{6}{2}$$

$$x - 1 = 3$$

$$x = 3 + 1$$

$$x = 4$$

26.

$$-3\Big[5 - 3\,(x + 4)\Big] = 12$$

$$5 - 3\,(x + 4) = \frac{12}{-3}$$

$$5 - 3\,(x + 4) = -4$$

$$-3\,(x + 4) = -4 - 5$$

$$-3\,(x + 4) = -9$$

$$x + 4 = \frac{-9}{-3}$$

$$x + 4 = 3$$

$$x = 3 - 4$$

$$x = -1$$

27.

$$2\Big[3 - (2x + 1)\Big] = 1$$

$$3 - (2x + 1) = \frac{1}{2}$$

$$-(2x + 1) = \frac{1}{2} - 3$$

$$-(2x + 1) = \frac{1}{2} - \frac{3}{1}$$

$$-(2x + 1) = \frac{1 - 6}{2}$$

$$-(2x + 1) = \frac{-5}{2}$$

$$2x + 1 = \frac{5}{2}$$

$$2x = \frac{5}{2} - 1$$

$$2x = \frac{5}{2} - \frac{1}{1}$$

$$2x = \frac{5 - 2}{2}$$

$$2x = \frac{3}{2}$$

$$x = \frac{1}{2} \times \frac{3}{2}$$

$$x = \frac{3}{4}$$

28.

$$-2\Big[5\,(x - 3) + 4\Big] = 3$$

$$5\,(x - 3) + 4 = \frac{3}{-2}$$

$$5\,(x - 3) + 4 = -\frac{3}{2}$$

$$5\,(x - 3) = -\frac{3}{2} - 4$$

$$5\,(x - 3) = \frac{-3}{2} - \frac{4}{1}$$

$$5\,(x - 3) = \frac{-3 - 8}{2}$$

$$5\,(x - 3) = \frac{-11}{2}$$

$$x - 3 = \frac{1}{5} \times \frac{-11}{2}$$

$$x - 3 = \frac{-11}{10}$$

$$x = \frac{-11}{10} + 3$$

$$x = \frac{-11}{10} + \frac{3}{1}$$

$$x = \frac{-11 + 30}{10}$$

$$x = \frac{19}{10}$$

Exercise Set 11.2D

1.

$$\frac{x}{5} = 2$$
$$x = 5 \times 2$$
$$x = 10$$

2.

$$\frac{x}{2} = -3$$
$$x = 2(-3)$$
$$x = -6$$

3.

$$\frac{x}{-3} = 4$$
$$x = -3 \times 4$$
$$x = -12$$

4.

$$\frac{y}{-5} = -1$$
$$y = -5(-1)$$
$$y = 5$$

5.

$$\frac{3}{x} = -4$$
$$x = \frac{3}{-4}$$
$$x = -\frac{3}{4}$$

6.

$$\frac{2}{y} = 5$$
$$y = \frac{2}{5}$$

7.

$$\frac{-5}{z} = \frac{2}{3}$$
$$z = \frac{3}{2}(-5)$$
$$z = -\frac{15}{2}$$

8.

$$\frac{4x}{5} = \frac{1}{6}$$
$$x = \frac{5}{4} \times \frac{1}{6}$$
$$x = \frac{5}{24}$$

9.

$$\frac{-x}{2} = -3$$
$$x = -2(-3)$$
$$x = 6$$

10.

$$\frac{3y}{-2} = 0$$
$$y = \frac{-2}{3} \times 0$$
$$y = 0$$

11.

$$\frac{2}{-3y} = 0$$
$$y = \frac{1}{0} \times \frac{2}{-3}$$

y is undefined.

No solutions.

12.

$$\frac{4}{5x} = -2$$
$$x = \frac{1}{-2} \times \frac{4}{5}$$
$$x = -\frac{2}{5}$$

13.

$$\frac{5}{2z} = \frac{1}{4}$$
$$z = \frac{4}{1} \times \frac{5}{2}$$
$$z = \frac{10}{1}$$
$$z = 10$$

14.

$$\frac{3}{-x} = -\frac{1}{3}$$

$$x = -\frac{3}{1} \times \frac{3}{-1}$$

$$x = \frac{9}{1}$$

$$x = 9$$

15.

$$\frac{x + 2}{2} = -3$$

$$x + 2 = 2(-3)$$

$$x + 2 = -6$$

$$x = -6 - 2$$

$$x = -8$$

16.

$$\frac{2x - 1}{3} = 5$$

$$2x - 1 = 3 \times 5$$

$$2x - 1 = 15$$

$$2x = 15 + 1$$

$$2x = 16$$

$$x = \frac{16}{2}$$

$$x = 8$$

17.

$$\frac{1 - y}{2} = \frac{1}{2}$$

$$1 - y = 2 \times \frac{1}{2}$$

$$1 - y = 1$$

$$-y = 1 - 1$$

$$-y = 0$$

$$y = 0$$

18.

$$\frac{3}{x - 3} = 4$$

$$x - 3 = \frac{3}{4}$$

$$x = \frac{3}{4} + 3$$

$$x = \frac{3}{4} + \frac{3}{1}$$

$$x = \frac{3 + 12}{4}$$

$$x = \frac{15}{4}$$

19.

$$\frac{-1}{2 - 3y} = -5$$

$$2 - 3y = \frac{-1}{-5}$$

$$2 - 3y = \frac{1}{5}$$

$$-3y = \frac{1}{5} - 2$$

$$-3y = \frac{1}{5} - \frac{2}{1}$$

$$-3y = \frac{1 - 10}{5}$$

$$-3y = \frac{-9}{5}$$

$$y = \frac{1}{-3} \times \frac{-9}{5}$$

$$y = \frac{3}{5}$$

20.

$$\frac{4}{3 - 2z} = \frac{3}{5}$$

$$3 - 2z = \frac{5}{3} \times \frac{4}{1}$$

$$3 - 2z = \frac{20}{3}$$

$$- 2z = \frac{20}{3} - 3$$

$$- 2z = \frac{20}{3} - \frac{3}{1}$$

$$- 2z = \frac{20 - 9}{3}$$

$$- 2z = \frac{11}{3}$$

$$z = -\frac{1}{2} \times \frac{11}{3}$$

$$z = -\frac{11}{6}$$

21.

$$\frac{1}{2 - 5z} = -1$$

$$2 - 5z = \frac{1}{-1}$$

$$2 - 5z = -1$$

$$- 5z = -1 - 2$$

$$- 5z = -3$$

$$z = \frac{-3}{-5}$$

$$z = \frac{3}{5}$$

Exercise Set 11.2E

1.

$$x^2 = 16$$

$$x = \pm\sqrt{16}$$

$$x = \pm 4$$

2.

$$x^2 = 25$$

$$x = \pm\sqrt{25}$$

$$x = \pm 5$$

3.

$$x^2 = 11$$

$$x = \pm\sqrt{11}$$

4.

$$x^2 = -5$$

$$x = \pm\sqrt{-5}$$

x is undefined.[12]
No solutions.

5.

$$y^3 = 125$$

$$y = \sqrt[3]{125}$$

$$y = 5$$

6.

$$y^3 = 1$$

$$y = \sqrt[3]{1}$$

$$y = 1$$

7.

$$y^3 = -64$$

$$y = \sqrt[3]{-64}$$

$$y = -4$$

[12]In this textbook we limit ourselves to the set of real numebrs, $\mathbb{R}$. The given equation does have solutions in the set of complex numbers, $\mathbb{C}$.

8.

$$z^2 + 3 = 4$$
$$z^2 = 4 - 3$$
$$z^2 = 1$$
$$z = \pm\sqrt{1}$$
$$z = \pm 1$$

9.

$$4z^3 + 1 = -3$$
$$4z^3 = -3 - 1$$
$$4z^3 = -4$$
$$z^3 = \frac{-4}{4}$$
$$z^3 = -1$$
$$z = \sqrt[3]{-1}$$
$$z = -1$$

10.

$$2z^3 - 2 = -4$$
$$2z^3 = -4 + 2$$
$$2z^3 = -2$$
$$z^3 = \frac{-2}{2}$$
$$z^3 = -1$$
$$z = \sqrt[3]{-1}$$
$$z = -1$$

11.

$$4\left(x^3 + 1\right) = 36$$
$$x^3 + 1 = \frac{36}{4}$$
$$x^3 + 1 = 9$$
$$x^3 = 9 - 1$$
$$x^3 = 8$$
$$x = \sqrt[3]{8}$$
$$x = 2$$

12.

$$-\left(5 - x^4\right) = 76$$
$$5 - x^4 = -76$$
$$-x^4 = -76 - 5$$
$$-x^4 = -81$$
$$x^4 = 81$$
$$x = \pm\sqrt[4]{81}$$
$$x = \pm 3$$

13.

$$-\left(2x^4 - 1\right) = 1$$
$$2x^4 - 1 = -1$$
$$2x^4 = -1 + 1$$
$$2x^4 = 0$$
$$x^4 = \frac{0}{2}$$
$$x^4 = 0$$
$$x = \sqrt[4]{0}$$
$$x = 0$$

14.

$$4^x = 64$$
$$x = \log_4 64$$
$$x = 3$$

15.

$$3^x = 9$$
$$x = \log_3 9$$
$$x = 2$$

16.

$$4^x = \frac{1}{16}$$
$$x = \log_4 \frac{1}{16}$$
$$x = -2$$

17.

$$3^y = \frac{1}{81}$$
$$y = \log_3 \frac{1}{81}$$
$$y = -4$$

18.

$$5^y + 2 = 27$$
$$5^y = 27 - 2$$
$$5^y = 25$$
$$y = \log_5 25$$
$$y = 2$$

19.

$$-3 + 2^y = -2$$
$$2^y = -2 + 3$$
$$2^y = 1$$
$$y = \log_2 1$$
$$y = 0$$

20.

$$4\left(5^z + 1\right) = 4$$
$$5^z + 1 = \frac{4}{4}$$
$$5^z + 1 = 1$$
$$5^z = 1 - 1$$
$$5^z = 0$$
$$z = \log_5 0$$
z is undefined.
No solutions.

21.

$$-3\left(2 - 4^z\right) = 6$$
$$2 - 4^z = \frac{6}{-3}$$
$$2 - 4^z = -2$$
$$-4^z = -2 - 2$$
$$-4^z = -4$$
$$4^z = 4$$
$$z = \log_4 4$$
$$z = 1$$

22.

$$\log x = 3$$
$$x = 10^3$$
$$x = 1000$$

23.

$$\log x = -1$$
$$x = 10^{-1}$$
$$x = \frac{1}{10^1}$$
$$x = \frac{1}{10}$$

24.

$$\log_2 x = 5$$
$$x = 2^5$$
$$x = 32$$

25.

$$\log_4 y = -2$$
$$y = 4^{-2}$$
$$y = \frac{1}{4^2}$$
$$y = \frac{1}{16}$$

26.

$$4\log_3 y = 0$$
$$\log_3 y = \frac{0}{4}$$
$$\log_3 y = 0$$
$$y = 3^0$$
$$y = 1$$

27.

$$-\log_2 z = -3$$
$$\log_2 z = 3$$
$$z = 2^3$$
$$z = 8$$

28.

$$\log z + 3 = 5$$
$$\log z = 5 - 3$$
$$\log z = 2$$
$$z = 10^2$$
$$z = 100$$

29.

$$\log (x + 3) = 1$$
$$x + 3 = 10^1$$
$$x + 3 = 10$$
$$x = 10 - 3$$
$$x = 7$$

30.

$$-2\log (y + 1) = -2$$
$$\log (y + 1) = \frac{-2}{-2}$$
$$\log (y + 1) = 1$$
$$y + 1 = 10^1$$
$$y + 1 = 10$$
$$y = 10 - 1$$
$$y = 9$$

31.

$$-\log (z - 2) = -1$$
$$\log (z - 2) = 1$$
$$z - 2 = 10^1$$
$$z - 2 = 10$$
$$z = 10 + 2$$
$$z = 12$$

32.

$$\frac{\log x}{4} = 1$$
$$\log x = 4 \times 1$$
$$\log x = 4$$
$$x = 10^4$$
$$x = 10\,000$$

33.

$$\log (x - 52) = 2$$
$$x - 52 = 10^2$$
$$x - 52 = 100$$
$$x = 100 + 52$$
$$x = 152$$

Exercise Set 11.2F

1.

$$2 (x + 1) + 3 (x - 2) = 6$$
$$2x + 2 + 3x - 6 = 6$$
$$5x - 4 = 6$$
$$5x = 6 + 4$$
$$5x = 10$$
$$x = \frac{10}{5}$$
$$x = 2$$

2.

$$5 (x + 2) - 2 (x + 5) = -3$$
$$5x + 10 - 2x - 10 = -3$$
$$3x = -3$$
$$x = \frac{-3}{3}$$
$$x = -1$$

3.

$$-3(x + 4) + 2x = -12$$
$$-3x - 12 + 2x = -12$$
$$-x - 12 = -12$$
$$-x = -12 + 12$$
$$-x = 0$$
$$x = 0$$

7.

$$3(2x + 5) = x - 3$$
$$6x + 15 = x - 3$$
$$6x - x = -3 - 15$$
$$5x = -18$$
$$x = \frac{-18}{5}$$
$$x = -\frac{18}{5}$$

4.

$$-(x - 1) - 4x = -14$$
$$-x + 1 - 4x = -14$$
$$-5x + 1 = -14$$
$$-5x = -14 - 1$$
$$-5x = -15$$
$$x = \frac{-15}{-5}$$
$$x = 3$$

8.

$$-(x + 4) = 3(x - 1)$$
$$-x - 4 = 3x - 3$$
$$-x - 3x = -3 + 4$$
$$-4x = 1$$
$$x = \frac{1}{-4}$$
$$x = -\frac{1}{4}$$

5.

$$5(x - 3) = 2(x + 3)$$
$$5x - 15 = 2x + 6$$
$$5x - 2x = 6 + 15$$
$$3x = 21$$
$$x = \frac{21}{3}$$
$$x = 7$$

9.

$$4x = -2(x + 6)$$
$$4x = -2x - 12$$
$$4x + 2x = -12$$
$$6x = -12$$
$$x = \frac{-12}{6}$$
$$x = -2$$

6.

$$4(5 - 2x) = -8x + 20$$
$$20 - 8x = -8x + 20$$
$$-8x + 8x = 20 - 20$$
$$0x = 0$$
$$x = \frac{0}{0}$$

x is indeterminate.

An infinite number of solutions.

$$x \in \mathbb{R}$$

10.

$$3(2x - 1) = 6x$$
$$6x - 3 = 6x$$
$$6x - 6x = -3$$
$$0x = -3$$
$$x = \frac{-3}{0}$$

x is undefined.

No solutions.

11.

$$8\left(2x + 1\right) - 3\left(x - 2\right) = -x$$
$$16x + 8 - 3x + 6 = -x$$
$$13x + 14 = -x$$
$$13x + x = -14$$
$$14x = -14$$
$$x = \frac{-14}{14}$$
$$x = -1$$

12.

$$4\left(x^3 + 3\right) - 2x^3 = 28$$
$$4x^3 + 12 - 2x^3 = 28$$
$$2x^3 + 12 = 28$$
$$2x^3 = 28 - 12$$
$$2x^3 = 16$$
$$x^3 = \frac{16}{2}$$
$$x^3 = 8$$
$$x = \sqrt[3]{8}$$
$$x = 2$$

13.

$$3x^2 - 2\left(x^2 + 5\right) = -9$$
$$3x^2 - 2x^2 - 10 = -9$$
$$x^2 - 10 = -9$$
$$x^2 = -9 + 10$$
$$x^2 = 1$$
$$x = \pm\sqrt{1}$$
$$x = \pm 1$$

14.

$$9\left(x^2 - 1\right) + 3\left(x^2 - 1\right) = -12$$
$$9x^2 - 9 + 3x^2 - 3 = -12$$
$$12x^2 - 12 = -12$$
$$12x^2 = -12 + 12$$
$$12x^2 = 0$$
$$x^2 = \frac{0}{12}$$
$$x^2 = 0$$
$$x = \sqrt{0}$$
$$x = 0$$

Exercise Set 11.2G

1.

$$-2 + 4x = 6$$
$$4x = 6 + 2$$
$$x = \frac{6 + 2}{4}$$
$$x = \frac{8}{4}$$
$$x = 2$$

2.

$$3x - 5 = 1$$
$$3x = 1 + 5$$
$$x = \frac{1 + 5}{3}$$
$$x = \frac{6}{3}$$
$$x = 2$$

3.

$$-2x - 6 = -14$$
$$-2x = -14 + 6$$
$$x = \frac{-14 + 6}{-2}$$
$$x = \frac{-8}{-2}$$
$$x = 4$$

4.

$$4 = 3 + 2x$$
$$3 + 2x = 4$$
$$2x = 4 - 3$$
$$x = \frac{4 - 3}{2}$$
$$x = \frac{1}{2}$$

5.

$$5 = -x + 3$$
$$-x + 3 = 5$$
$$-x = 5 - 3$$
$$x = -(5 - 3)$$
$$x = -2$$

6.

$$-7 + 8x = 0$$
$$8x = 0 + 7$$
$$x = \frac{0 + 7}{8}$$
$$x = \frac{7}{8}$$

7.

$$9 - 2x = 2$$
$$-2x = 2 - 9$$
$$x = \frac{2 - 9}{-2}$$
$$x = \frac{-7}{-2}$$
$$x = \frac{7}{2}$$

8.

$$6x + 1 = -3$$
$$6x = -3 - 1$$
$$x = \frac{-3 - 1}{6}$$
$$x = \frac{-4}{6}$$
$$x = -\frac{2}{3}$$

9.

$$-8x - 6 = -7$$
$$-8x = -7 + 6$$
$$x = \frac{-7 + 6}{-8}$$
$$x = \frac{-1}{-8}$$
$$x = \frac{1}{8}$$

Exercise Set 11.2H

1.

$$\frac{3}{5}x + \frac{1}{2} = \frac{1}{3}$$
$$\frac{3}{5}x = \frac{1}{3} - \frac{1}{2}$$
$$x = \frac{5}{3}\left(\frac{1}{3} - \frac{1}{2}\right)$$
$$x = \frac{5}{3} \times \frac{2 - 3}{6}$$
$$x = \frac{5}{3} \times \frac{-1}{6}$$
$$x = -\frac{5}{18}$$

2.

$$\frac{4}{7}x + \frac{3}{4} = \frac{2}{5}$$
$$\frac{4}{7}x = \frac{2}{5} - \frac{3}{4}$$
$$x = \frac{7}{4}\left(\frac{2}{5} - \frac{3}{4}\right)$$
$$x = \frac{7}{4} \times \frac{8 - 15}{20}$$
$$x = \frac{7}{4} \times \frac{-7}{20}$$
$$x = -\frac{49}{80}$$

5.

$$-\frac{4}{9} + 2x = \frac{1}{3}$$
$$2x = \frac{1}{3} + \frac{4}{9}$$
$$x = \frac{1}{2}\left(\frac{1}{3} + \frac{4}{9}\right)$$
$$x = \frac{1}{2} \times \frac{3 + 4}{9}$$
$$x = \frac{1}{2} \times \frac{7}{9}$$
$$x = \frac{7}{18}$$

3.

$$-\frac{1}{3}x - \frac{2}{9} = -\frac{5}{6}$$
$$-\frac{1}{3}x = -\frac{5}{6} + \frac{2}{9}$$
$$x = -\frac{3}{1}\left(-\frac{5}{6} + \frac{2}{9}\right)$$
$$x = -\frac{3}{1}\left(\frac{-5}{6} + \frac{2}{9}\right)$$
$$x = -\frac{3}{1} \times \frac{-15 + 4}{18}$$
$$x = -\frac{3}{1} \times \frac{-11}{18}$$
$$x = \frac{11}{6}$$

6.

$$-4x - \frac{2}{5} = -\frac{1}{4}$$
$$-4x = -\frac{1}{4} + \frac{2}{5}$$
$$x = -\frac{1}{4}\left(-\frac{1}{4} + \frac{2}{5}\right)$$
$$x = -\frac{1}{4}\left(\frac{-1}{4} + \frac{2}{5}\right)$$
$$x = -\frac{1}{4} \times \frac{-5 + 8}{20}$$
$$x = -\frac{1}{4} \times \frac{3}{20}$$
$$x = -\frac{3}{80}$$

4.

$$\frac{1}{8} + \frac{3}{5}x = \frac{1}{2}$$
$$\frac{3}{5}x = \frac{1}{2} - \frac{1}{8}$$
$$x = \frac{5}{3}\left(\frac{1}{2} - \frac{1}{8}\right)$$
$$x = \frac{5}{3} \times \frac{4 - 1}{8}$$
$$x = \frac{5}{3} \times \frac{3}{8}$$
$$x = \frac{5}{8}$$

7.

$$-x + 1 = \frac{3}{4}$$
$$-x = \frac{3}{4} - 1$$
$$x = -\frac{3}{4} + 1$$
$$x = \frac{-3}{4} + \frac{1}{1}$$
$$x = \frac{-3 + 4}{4}$$
$$x = \frac{1}{4}$$

8.

$$2x - 3 = \frac{4}{5}$$

$$2x = \frac{4}{5} + 3$$

$$x = \frac{1}{2}\left(\frac{4}{5} + 3\right)$$

$$x = \frac{1}{2}\left(\frac{4}{5} + \frac{3}{1}\right)$$

$$x = \frac{1}{2} \times \frac{4 + 15}{5}$$

$$x = \frac{1}{2} \times \frac{19}{5}$$

$$x = \frac{19}{10}$$

9.

$$3 - 5x = \frac{9}{2}$$

$$-5x = \frac{9}{2} - 3$$

$$x = -\frac{1}{5}\left(\frac{9}{2} - 3\right)$$

$$x = -\frac{1}{5}\left(\frac{9}{2} - \frac{3}{1}\right)$$

$$x = -\frac{1}{5} \times \frac{9 - 6}{2}$$

$$x = -\frac{1}{5} \times \frac{3}{2}$$

$$x = -\frac{3}{10}$$

10.

$$-2 + \frac{2}{3}x = \frac{1}{4}$$

$$\frac{2}{3}x = \frac{1}{4} + 2$$

$$x = \frac{3}{2}\left(\frac{1}{4} + 2\right)$$

$$x = \frac{3}{2}\left(\frac{1}{4} + \frac{2}{1}\right)$$

$$x = \frac{3}{2} \times \frac{1 + 8}{4}$$

$$x = \frac{3}{2} \times \frac{9}{4}$$

$$x = \frac{27}{8}$$

11.

$$-\frac{5}{6}x - \frac{3}{4} = 1$$

$$-\frac{5}{6}x = 1 + \frac{3}{4}$$

$$x = -\frac{6}{5}\left(1 + \frac{3}{4}\right)$$

$$x = -\frac{6}{5}\left(\frac{1}{1} + \frac{3}{4}\right)$$

$$x = -\frac{6}{5} \times \frac{4 + 3}{4}$$

$$x = -\frac{6}{5} \times \frac{7}{4}$$

$$x = -\frac{21}{10}$$

12.

$$\frac{8}{3}x - \frac{8}{3} = 2$$

$$\frac{8}{3}x = 2 + \frac{8}{3}$$

$$x = \frac{3}{8}\left(2 + \frac{8}{3}\right)$$

$$x = \frac{3}{8}\left(\frac{2}{1} + \frac{8}{3}\right)$$

$$x = \frac{3}{8} \times \frac{6 + 8}{3}$$

$$x = \frac{3}{8} \times \frac{14}{3}$$

$$x = \frac{7}{4}$$

13.

$$\frac{1}{3} = -2 + \frac{4}{7}x$$

$$-2 + \frac{4}{7}x = \frac{1}{3}$$

$$\frac{4}{7}x = \frac{1}{3} + 2$$

$$x = \frac{7}{4}\left(\frac{1}{3} + 2\right)$$

$$x = \frac{7}{4}\left(\frac{1}{3} + \frac{2}{1}\right)$$

$$x = \frac{7}{4} \times \frac{1 + 6}{3}$$

$$x = \frac{7}{4} \times \frac{7}{3}$$

$$x = \frac{49}{12}$$

14.

$$-3 = -\frac{7}{5}x + 2$$

$$-\frac{7}{5}x + 2 = -3$$

$$-\frac{7}{5}x = -3 - 2$$

$$x = -\frac{5}{7}(-3 - 2)$$

$$x = -\frac{5}{7}(-5)$$

$$x = -\frac{5}{7} \times \frac{-5}{1}$$

$$x = \frac{25}{7}$$

Exercise Set 11.2I

1.

$$4(x + 3) = 20$$

$$x + 3 = \frac{20}{4}$$

$$x = \frac{20}{4} - 3$$

$$x = 5 - 3$$

$$x = 2$$

3.

$$2(8 + x) = 18$$

$$8 + x = \frac{18}{2}$$

$$x = \frac{18}{2} - 8$$

$$x = 9 - 8$$

$$x = 1$$

2.

$$3(x + 5) = 24$$

$$x + 5 = \frac{24}{3}$$

$$x = \frac{24}{3} - 5$$

$$x = 8 - 5$$

$$x = 3$$

4.

$$3(2 + x) = 6$$

$$2 + x = \frac{6}{3}$$

$$x = \frac{6}{3} - 2$$

$$x = 2 - 2$$

$$x = 0$$

5.

$$5\left(x - 4\right) = 10$$
$$x - 4 = \frac{10}{5}$$
$$x = \frac{10}{5} + 4$$
$$x = 2 + 4$$
$$x = 6$$

9.

$$-2\left(y - 5\right) = 8$$
$$y - 5 = \frac{8}{-2}$$
$$y = \frac{8}{-2} + 5$$
$$y = -4 + 5$$
$$y = 1$$

6.

$$7\left(x - 2\right) = -7$$
$$x - 2 = \frac{-7}{7}$$
$$x = \frac{-7}{7} + 2$$
$$x = -1 + 2$$
$$x = 1$$

10.

$$-3\left(4 + y\right) = -21$$
$$4 + y = \frac{-21}{-3}$$
$$y = \frac{-21}{-3} - 4$$
$$y = 7 - 4$$
$$y = 3$$

7.

$$32 = 8\left(5 - y\right)$$
$$8\left(5 - y\right) = 32$$
$$5 - y = \frac{32}{8}$$
$$-y = \frac{32}{8} - 5$$
$$y = -\left(\frac{32}{8} - 5\right)$$
$$y = -\left(4 - 5\right)$$
$$y = -\left(-1\right)$$
$$y = 1$$

11.

$$-2\left(5 - y\right) = -6$$
$$5 - y = \frac{-6}{-2}$$
$$-y = \frac{-6}{-2} - 5$$
$$y = -\left(\frac{-6}{-2} - 5\right)$$
$$y = -\left(3 - 5\right)$$
$$y = -\left(-2\right)$$
$$y = 2$$

8.

$$18 = 3\left(2 - y\right)$$
$$3\left(2 - y\right) = 18$$
$$2 - y = \frac{18}{3}$$
$$-y = \frac{18}{3} - 2$$
$$y = -\left(\frac{18}{3} - 2\right)$$
$$y = -\left(6 - 2\right)$$
$$y = -4$$

12.

$$-\left(2y + 3\right) = -3$$
$$2y + 3 = -\left(-3\right)$$
$$2y = -\left(-3\right) - 3$$
$$y = \frac{-\left(-3\right) - 3}{2}$$
$$y = \frac{3 - 3}{2}$$
$$y = \frac{0}{2}$$
$$y = 0$$

13.

$$-(4 - 5z) = 6$$
$$4 - 5z = -6$$
$$-5z = -6 - 4$$
$$z = \frac{-6 - 4}{-5}$$
$$z = \frac{-10}{-5}$$
$$z = 2$$

14.

$$5(3z - 1) = -5$$
$$3z - 1 = \frac{-5}{5}$$
$$3z = \frac{-5}{5} + 1$$
$$z = \frac{1}{3}\left(\frac{-5}{5} + 1\right)$$
$$z = \frac{1}{3}(-1 + 1)$$
$$z = \frac{1}{3} \times 0$$
$$z = 0$$

15.

$$-2(5 - 3z) = 8$$
$$5 - 3z = \frac{8}{-2}$$
$$-3z = \frac{8}{-2} - 5$$
$$z = -\frac{1}{3}\left(\frac{8}{-2} - 5\right)$$
$$z = -\frac{1}{3}(-4 - 5)$$
$$z = -\frac{1}{3}(-9)$$
$$z = 3$$

16.

$$3(4z + 2) = 7$$
$$4z + 2 = \frac{7}{3}$$
$$4z = \frac{7}{3} - 2$$
$$z = \frac{1}{4}\left(\frac{7}{3} - 2\right)$$
$$z = \frac{1}{4}\left(\frac{7}{3} - \frac{2}{1}\right)$$
$$z = \frac{1}{4} \times \frac{7 - 6}{3}$$
$$z = \frac{1}{4} \times \frac{1}{3}$$
$$z = \frac{1}{12}$$

17.

$$-2(-z + 1) = 5$$
$$-z + 1 = \frac{5}{-2}$$
$$-z = \frac{5}{-2} - 1$$
$$z = -\left(\frac{5}{-2} - 1\right)$$
$$z = -\left(\frac{-5}{2} - \frac{1}{1}\right)$$
$$z = -\frac{-5 - 2}{2}$$
$$z = -\frac{-7}{2}$$
$$z = \frac{7}{2}$$

18.

$$\frac{1}{2}(2x + 4) = 3$$
$$2x + 4 = 2 \times 3$$
$$2x = 2 \times 3 - 4$$
$$x = \frac{2 \times 3 - 4}{2}$$
$$x = \frac{6 - 4}{2}$$
$$x = \frac{2}{2}$$
$$x = 1$$

19.

$$\frac{2}{3}\left(\frac{3}{4}x - \frac{1}{2}\right) = \frac{1}{3}$$

$$\frac{3}{4}x - \frac{1}{2} = \frac{3}{2} \times \frac{1}{3}$$

$$\frac{3}{4}x = \frac{3}{2} \times \frac{1}{3} + \frac{1}{2}$$

$$x = \frac{4}{3}\left(\frac{3}{2} \times \frac{1}{3} + \frac{1}{2}\right)$$

$$x = \frac{4}{3}\left(\frac{1}{2} + \frac{1}{2}\right)$$

$$x = \frac{4}{3} \times \frac{1 + 1}{2}$$

$$x = \frac{4}{3} \times \frac{2}{2}$$

$$x = \frac{4}{3}$$

20.

$$\frac{3}{4}\left(\frac{5}{3}y - 1\right) = \frac{1}{3}$$

$$\frac{5}{3}y - 1 = \frac{4}{3} \times \frac{1}{3}$$

$$\frac{5}{3}y = \frac{4}{3} \times \frac{1}{3} + 1$$

$$x = \frac{3}{5}\left(\frac{4}{3} \times \frac{1}{3} + 1\right)$$

$$x = \frac{3}{5}\left(\frac{4}{9} + 1\right)$$

$$x = \frac{3}{5}\left(\frac{4}{9} + \frac{1}{1}\right)$$

$$x = \frac{3}{5} \times \frac{4 + 9}{9}$$

$$x = \frac{3}{5} \times \frac{13}{9}$$

$$x = \frac{13}{15}$$

21.

$$-\frac{2}{3}\left(z - \frac{3}{4}\right) = \frac{1}{5}$$

$$z - \frac{3}{4} = -\frac{3}{2} \times \frac{1}{5}$$

$$z = -\frac{3}{2} \times \frac{1}{5} + \frac{3}{4}$$

$$z = -\frac{3}{10} + \frac{3}{4}$$

$$z = \frac{-3}{10} + \frac{3}{4}$$

$$z = \frac{-6 + 15}{20}$$

$$z = \frac{9}{20}$$

22.

$$4(x + 2) - 1 = 19$$

$$4(x + 2) = 19 + 1$$

$$x + 2 = \frac{19 + 1}{4}$$

$$x = \frac{19 + 1}{4} - 2$$

$$x = \frac{20}{4} - 2$$

$$x = 5 - 2$$

$$x = 3$$

23.

$$3(2x + 1) + 4 = 1$$

$$3(2x + 1) = 1 - 4$$

$$2x + 1 = \frac{1 - 4}{3}$$

$$2x = \frac{1 - 4}{3} - 1$$

$$x = \frac{1}{2}\left(\frac{1 - 4}{3} - 1\right)$$

$$x = \frac{1}{2}\left(\frac{-3}{3} - 1\right)$$

$$x = \frac{1}{2}(-1 - 1)$$

$$x = \frac{1}{2}(-2)$$

$$x = -1$$

24.

$$-(3x - 1) = 0$$
$$3x - 1 = 0$$
$$3x = 0 + 1$$
$$x = \frac{0 + 1}{3}$$
$$x = \frac{1}{3}$$

25.

$$5\left[2(x - 1) + 3\right] = 45$$
$$2(x - 1) + 3 = \frac{45}{5}$$
$$2(x - 1) = \frac{45}{5} - 3$$
$$x - 1 = \frac{1}{2}\left(\frac{45}{5} - 3\right)$$
$$x = \frac{1}{2}\left(\frac{45}{5} - 3\right) + 1$$
$$x = \frac{1}{2}(9 - 3) + 1$$
$$x = \frac{1}{2} \times 6 + 1$$
$$x = 3 + 1$$
$$x = 4$$

26.

$$-3\left[5 - 3(x + 4)\right] = 12$$
$$5 - 3(x + 4) = \frac{12}{-3}$$
$$-3(x + 4) = \frac{12}{-3} - 5$$
$$x + 4 = \frac{1}{-3}\left(\frac{12}{-3} - 5\right)$$
$$x = \frac{1}{-3}\left(\frac{12}{-3} - 5\right) - 4$$
$$x = \frac{1}{-3}(-4 - 5) - 4$$
$$x = \frac{1}{-3}(-9) - 4$$
$$x = 3 - 4$$
$$x = -1$$

27.

$$2\left[3 - (2x + 1)\right] = 1$$
$$3 - (2x + 1) = \frac{1}{2}$$
$$-(2x + 1) = \frac{1}{2} - 3$$
$$2x + 1 = -\left(\frac{1}{2} - 3\right)$$
$$2x = -\left(\frac{1}{2} - 3\right) - 1$$
$$x = \frac{1}{2}\left[-\left(\frac{1}{2} - 3\right) - 1\right]$$
$$x = \frac{1}{2}\left[-\left(\frac{1}{2} - \frac{3}{1}\right) - 1\right]$$
$$x = \frac{1}{2}\left[-\left(\frac{1 - 6}{2}\right) - 1\right]$$
$$x = \frac{1}{2}\left[-\left(\frac{-5}{2}\right) - 1\right]$$
$$x = \frac{1}{2}\left(\frac{5}{2} - 1\right)$$
$$x = \frac{1}{2}\left(\frac{5}{2} - \frac{1}{1}\right)$$
$$x = \frac{1}{2}\left(\frac{5 - 2}{2}\right)$$
$$x = \frac{1}{2} \times \frac{3}{2}$$
$$x = \frac{3}{4}$$

28.

$$-2\left[5(x - 3) + 4\right] = 3$$

$$5(x - 3) + 4 = \frac{3}{-2}$$

$$5(x - 3) = \frac{3}{-2} - 4$$

$$x - 3 = \frac{1}{5}\left(\frac{3}{-2} - 4\right)$$

$$x = \frac{1}{5}\left(\frac{3}{-2} - 4\right) + 3$$

$$x = \frac{1}{5}\left(\frac{-3}{2} - \frac{4}{1}\right) + 3$$

$$x = \frac{1}{5} \times \frac{-3 - 8}{2} + 3$$

$$x = \frac{1}{5} \times \frac{-11}{2} + 3$$

$$x = -\frac{11}{10} + 3$$

$$x = \frac{-11}{10} + \frac{3}{1}$$

$$x = \frac{-11 + 30}{10}$$

$$x = \frac{19}{10}$$

Exercise Set 11.2J

1.

$$\frac{x + 2}{2} = -3$$

$$x + 2 = 2(-3)$$

$$x = 2(-3) - 2$$

$$x = -6 - 2$$

$$x = -8$$

2.

$$\frac{2x - 1}{3} = 5$$

$$2x - 1 = 3 \times 5$$

$$2x = 3 \times 5 + 1$$

$$x = \frac{3 \times 5 + 1}{2}$$

$$x = \frac{15 + 1}{2}$$

$$x = \frac{16}{2}$$

$$x = 8$$

3.

$$\frac{1-y}{2} = \frac{1}{2}$$

$$1 - y = 2 \times \frac{1}{2}$$

$$-y = 2 \times \frac{1}{2} - 1$$

$$y = -\left(2 \times \frac{1}{2} - 1\right)$$

$$x = -(1 - 1)$$

$$x = 0$$

4.

$$\frac{3}{x-3} = 4$$

$$x - 3 = \frac{3}{4}$$

$$x = \frac{3}{4} + 3$$

$$x = \frac{3}{4} + \frac{3}{1}$$

$$x = \frac{3 + 12}{4}$$

$$x = \frac{15}{4}$$

5.

$$\frac{-1}{2 - 3y} = -5$$

$$2 - 3y = \frac{-1}{-5}$$

$$-3y = \frac{-1}{-5} - 2$$

$$x = -\frac{1}{3}\left(\frac{-1}{-5} - 2\right)$$

$$x = -\frac{1}{3}\left(\frac{1}{5} - \frac{2}{1}\right)$$

$$x = -\frac{1}{3} \times \frac{1 - 10}{5}$$

$$x = -\frac{1}{3} \times \frac{-9}{5}$$

$$x = \frac{3}{5}$$

6.

$$\frac{4}{3 - 2z} = \frac{3}{5}$$

$$3 - 2z = \frac{5}{3} \times 4$$

$$-2z = \frac{5}{3} \times 4 - 3$$

$$x = -\frac{1}{2}\left(\frac{5}{3} \times 4 - 3\right)$$

$$x = -\frac{1}{2}\left(\frac{20}{3} - \frac{3}{1}\right)$$

$$x = -\frac{1}{2} \times \frac{20 - 9}{3}$$

$$x = -\frac{1}{2} \times \frac{11}{3}$$

$$x = -\frac{11}{6}$$

7.

$$\frac{1}{2 - 5z} = -1$$

$$\frac{1}{-1} = 2 - 5z$$

$$2 - 5z = \frac{1}{-1}$$

$$-5z = \frac{1}{-1} - 2$$

$$x = -\frac{1}{5}\left(\frac{1}{-1} - 2\right)$$

$$x = -\frac{1}{5}(-1 - 2)$$

$$x = -\frac{1}{5}(-3)$$

$$x = \frac{3}{5}$$

8.

$$\frac{3z + 1}{2 - 5z} = -1$$

$$3z + 1 = -(2 - 5z)$$

$$3z + 1 = -2 + 5z$$

$$3z - 5z = -2 - 1$$

$$(3 - 5)z = -2 - 1$$

$$z = \frac{-2 - 1}{3 - 5}$$

$$z = \frac{-3}{-2}$$

$$z = \frac{3}{2}$$

9.

$$\frac{4 - z}{z + 5} = 2$$

$$4 - z = 2(z + 5)$$

$$4 - z = 2z + 2 \times 5$$

$$-z - 2z = 2 \times 5 - 4$$

$$(-1 - 2)z = 2 \times 5 - 4$$

$$z = \frac{2 \times 5 - 4}{-1 - 2}$$

$$z = \frac{10 - 4}{-3}$$

$$z = \frac{6}{-3}$$

$$z = -2$$

Exercise Set 11.2K

1.

$$z^2 + 3 = 4$$

$$z^2 = 4 - 3$$

$$z = \pm\sqrt{4 - 3}$$

$$z = \pm\sqrt{1}$$

$$z = \pm 1$$

2.

$$4z^3 + 1 = -3$$

$$4z^3 = -3 - 1$$

$$z^3 = \frac{-3 - 1}{4}$$

$$z = \sqrt[3]{\frac{-3 - 1}{4}}$$

$$z = \sqrt[3]{\frac{-4}{4}}$$

$$z = \sqrt[3]{-1}$$

$$z = -1$$

3.

$$2z^3 - 2 = -4$$

$$2z^3 = -4 + 2$$

$$z^3 = \frac{-4 + 2}{2}$$

$$z = \sqrt[3]{\frac{-4 + 2}{2}}$$

$$z = \sqrt[3]{\frac{-2}{2}}$$

$$z = \sqrt[3]{-1}$$

$$z = -1$$

4.

$$4(x^3 + 1) = 36$$

$$x^3 + 1 = \frac{36}{4}$$

$$x^3 = \frac{36}{4} - 1$$

$$x = \sqrt[3]{\frac{36}{4} - 1}$$

$$x = \sqrt[3]{9 - 1}$$

$$x = \sqrt[3]{8}$$

$$x = 2$$

5.

$$-\left(5 - x^4\right) = 76$$
$$5 - x^4 = -76$$
$$-x^4 = -76 - 5$$
$$x^4 = -(-76 - 5)$$
$$x^4 = \pm\sqrt[4]{-(-76 - 5)}$$
$$x^4 = \pm\sqrt[4]{-(-81)}$$
$$x = \pm\sqrt[4]{81}$$
$$x = \pm 3$$

6.

$$-\left(2x^4 - 1\right) = 1$$
$$2x^4 - 1 = -1$$
$$2x^4 = -1 + 1$$
$$x^4 = \frac{-1 + 1}{2}$$
$$x = \pm\sqrt[4]{\frac{-1 + 1}{2}}$$
$$x = \pm\sqrt[4]{\frac{0}{2}}$$
$$x = \sqrt[4]{0}$$
$$x = 0$$

7.

$$5^y + 2 = 27$$
$$5^y = 27 - 2$$
$$y = \log_5\left(27 - 2\right)$$
$$y = \log_5 25$$
$$y = 2$$

8.

$$-3 + 2^y = -2$$
$$2^y = -2 + 3$$
$$y = \log_2\left(-2 + 3\right)$$
$$y = \log_2 1$$
$$y = 0$$

9.

$$4\left(5^z + 1\right) = 4$$
$$5^z + 1 = \frac{4}{4}$$
$$5^z = \frac{4}{4} - 1$$
$$z = \log_5\left(\frac{4}{4} - 1\right)$$
$$z = \log_5\left(1 - 1\right)$$
$$z = \log_5 0$$
z is undefined.
No solutions.

10.

$$-3\left(2 - 4^z\right) = 6$$
$$2 - 4^z = \frac{6}{-3}$$
$$-4^z = \frac{6}{-3} - 2$$
$$4^z = -\left(\frac{6}{-3} - 2\right)$$
$$z = \log_4 -\left(\frac{6}{-3} - 2\right)$$
$$z = \log_4 -(-2 - 2)$$
$$z = \log_4 -(-4)$$
$$z = \log_4 4$$
$$z = 1$$

11.

$$\log z + 3 = 5$$
$$\log z = 5 - 3$$
$$z = 10^{5-3}$$
$$z = 10^2$$
$$z = 100$$

12.

$$\log\left(x + 3\right) = 2$$
$$x + 3 = 10^2$$
$$x = 10^2 - 3$$
$$x = 100 - 3$$
$$x = 97$$

13.

$$-2\log y + 1 = -1$$
$$-2\log y = -1 - 1$$
$$\log y = \frac{-1 - 1}{-2}$$
$$y = 10^{\frac{-1-1}{-2}}$$
$$y = 10^{\frac{-2}{-2}}$$
$$y = 10^1$$
$$y = 10$$

14.

$$-\log(z - 2) = -1$$
$$\log(z - 2) = -(-1)$$
$$z - 2 = 10^{-(-1)}$$
$$z = 10^{-(-1)} + 2$$
$$z = 10^1 + 2$$
$$z = 10 + 2$$
$$z = 12$$

15.

$$\frac{\log x}{2} = 1$$
$$\log x = 2 \times 1$$
$$x = 10^{2 \times 1}$$
$$x = 10^2$$
$$x = 100$$

16.

$$\frac{4}{\log(x - 1)} = 2$$
$$\log(x - 1) = \frac{4}{2}$$
$$x - 1 = 10^{\frac{4}{2}}$$
$$x = 10^{\frac{4}{2}} + 1$$
$$x = 10^2 + 1$$
$$x = 100 + 1$$
$$x = 101$$

Exercise Set 11.2L

1.

$$2(x + 1) + 3(x - 2) = 6$$
$$2x + 2 \times 1 + 3x - 3 \times 2 = 6$$
$$2x + 3x = 6 - 2 \times 1 + 3 \times 2$$
$$(2 + 3)x = 6 - 2 \times 1 + 3 \times 2$$
$$x = \frac{6 - 2 \times 1 + 3 \times 2}{2 + 3}$$
$$x = \frac{6 - 2 + 6}{5}$$
$$x = \frac{10}{5}$$
$$x = 2$$

2.

$$5(x + 2) - 2(x + 5) = -3$$
$$5x + 5 \times 2 - 2x - 2 \times 5 = -3$$
$$5x - 2x = -3 - 5 \times 2 + 2 \times 5$$
$$(5 - 2)x = -3 - 5 \times 2 + 2 \times 5$$
$$x = \frac{-3 - 5 \times 2 + 2 \times 5}{5 - 2}$$
$$x = \frac{-3 - 10 + 10}{3}$$
$$x = \frac{-3}{3}$$
$$x = -1$$

3.

$$-3(x + 4) + 2x = -12$$
$$-3x - 3 \times 4 + 2x = -12$$
$$-3x + 2x = -12 + 3 \times 4$$
$$(-3 + 2)x = -12 + 3 \times 4$$
$$x = \frac{-12 + 3 \times 4}{-3 + 2}$$
$$x = \frac{-12 + 12}{-1}$$
$$x = \frac{0}{-1}$$
$$x = 0$$

6.

$$4(5 - 2x) = -8x + 20$$
$$4 \times 5 - 4 \times 2x = -8x + 20$$
$$-4 \times 2x + 8x = 20 - 4 \times 5$$
$$(-4 \times 2 + 8)x = 20 - 4 \times 5$$
$$x = \frac{20 - 4 \times 5}{-4 \times 2 + 8}$$
$$x = \frac{20 - 20}{-8 + 8}$$
$$x = \frac{0}{0}$$

x is indeterminate.

An infinite number of solutions.

$x \in \mathbb{R}$

4.

$$-(x - 1) - 4x = -14$$
$$-x + 1 - 4x = -14$$
$$-x - 4x = -14 - 1$$
$$(-1 - 4)x = -14 - 1$$
$$x = \frac{-14 - 1}{-1 - 4}$$
$$x = \frac{-15}{-5}$$
$$x = 3$$

7.

$$3(2x + 5) = x - 3$$
$$3 \times 2x + 3 \times 5 = x - 3$$
$$3 \times 2x - x = -3 - 3 \times 5$$
$$(3 \times 2 - 1)x = -3 - 3 \times 5$$
$$x = \frac{-3 - 3 \times 5}{3 \times 2 - 1}$$
$$x = \frac{-3 - 15}{6 - 1}$$
$$x = \frac{-18}{5}$$
$$x = -\frac{18}{5}$$

5.

$$5(x - 3) = 2(x + 3)$$
$$5x - 5 \times 3 = 2x + 2 \times 3$$
$$5x - 2x = 2 \times 3 + 5 \times 3$$
$$(5 - 2)x = 2 \times 3 + 5 \times 3$$
$$x = \frac{2 \times 3 + 5 \times 3}{5 - 2}$$
$$x = \frac{6 + 15}{3}$$
$$x = \frac{21}{3}$$
$$x = 7$$

8.

$$-(x + 4) = 3(x - 1)$$
$$-x - 4 = 3x - 3 \times 1$$
$$-x - 3x = -3 \times 1 + 4$$
$$(-1 - 3)x = -3 \times 1 + 4$$
$$x = \frac{-3 \times 1 + 4}{-1 - 3}$$
$$x = \frac{-3 + 4}{-4}$$
$$x = \frac{1}{-4}$$
$$x = -\frac{1}{4}$$

9.

$$4x = -2(x + 6)$$
$$4x = -2x - 2 \times 6$$
$$4x + 2x = -2 \times 6$$
$$(4 + 2)x = -2 \times 6$$
$$x = \frac{-2 \times 6}{4 + 2}$$
$$x = \frac{-2 \times 6}{6}$$
$$x = -2$$

10.

$$3(2x - 1) = 6x$$
$$3 \times 2x - 3 \times 1 = 6x$$
$$3 \times 2x - 6x = 3 \times 1$$
$$(3 \times 2 - 6)x = 3 \times 1$$
$$x = \frac{3 \times 1}{3 \times 2 - 6}$$
$$x = \frac{3}{6 - 6}$$
$$x = \frac{3}{0}$$

x is undefined.

No solutions.

11.

$$8(2x + 1) - 3(x - 2) = -x$$
$$8 \times 2x + 8 \times 1 - 3x + 3 \times 2 = -x$$
$$8 \times 2x - 3x + x = -8 \times 1 - 3 \times 2$$
$$(8 \times 2 - 3 + 1)x = -8 \times 1 - 3 \times 2$$
$$x = \frac{-8 \times 1 - 3 \times 2}{8 \times 2 - 3 + 1}$$
$$x = \frac{-8 - 6}{16 - 3 + 1}$$
$$x = \frac{-14}{14}$$
$$x = -1$$

12.

$$4(x^3 + 3) - 2x^3 = 28$$
$$4x^3 + 4 \times 3 - 2x^3 = 28$$
$$4x^3 - 2x^3 = 28 - 4 \times 3$$
$$(4 - 2)x^3 = 28 - 4 \times 3$$
$$x^3 = \frac{28 - 4 \times 3}{4 - 2}$$
$$x = \sqrt[3]{\frac{28 - 4 \times 3}{4 - 2}}$$
$$x = \sqrt[3]{\frac{28 - 12}{2}}$$
$$x = \sqrt[3]{\frac{16}{2}}$$
$$x = \sqrt[3]{8}$$
$$x = 2$$

13.

$$3x^2 - 2(x^2 + 5) = -9$$
$$3x^2 - 2x^2 - 2 \times 5 = -9$$
$$3x^2 - 2x^2 = -9 + 2 \times 5$$
$$(3 - 2)x^2 = -9 + 2 \times 5$$
$$x^2 = \frac{-9 + 2 \times 5}{3 - 2}$$
$$x = \pm\sqrt{\frac{-9 + 2 \times 5}{3 - 2}}$$
$$x = \pm\sqrt{\frac{-9 + 10}{1}}$$
$$x = \pm\sqrt{\frac{1}{1}}$$
$$x = \pm\sqrt{1}$$
$$x = \pm 1$$

14.

$$9\left(x^2 - 1\right) + 3\left(x^2 - 1\right) = -12$$

$$9x^2 - 9 \times 1 + 3x^2 - 3 \times 1 = -12$$

$$9x^2 + 3x^2 = -12 + 9 \times 1 + 3 \times 1$$

$$(9 + 3)x^2 = -12 + 9 \times 1 + 3 \times 1$$

$$x^2 = \frac{-12 + 9 \times 1 + 3 \times 1}{9 + 3}$$

$$x = \pm\sqrt{\frac{-12 + 9 \times 1 + 3 \times 1}{9 + 3}}$$

$$x = \pm\sqrt{\frac{-12 + 9 + 3}{12}}$$

$$x = \pm\sqrt{\frac{0}{12}}$$

$$x = \sqrt{0}$$

$$x = 0$$

Exercise Set 12.2.1.1

1. It means that an increase or decrease in the value of one of the quantities forces a corresponding increase or decrease in the value of the other quantity.

2. It means that an increase or decrease in the value of one of the quantities by a certain amount forces a corresponding increase or decrease in the value of the other quantity by the same amount.

3. $y = k + x$, $k \in \mathbb{R}$

4. $y - x = k$, $k \in \mathbb{R}$

5. It means that an increase or decrease in the value of either quantity by a certain amount forces a corresponding increase or decrease in the value of the other quantity by the same amount.

6. $\Delta y = \Delta x$ where x and y represent the values of the two quantities.

7. The shape of the graph of the model for a relationship of type direct superposition is a straight line.

8. 1

9. $45\,^\circ$

10. The effect of k in $y = k + x$, $k \in \mathbb{R}$, is to raise or lower the line. The higher the value of k, the higher the line.

11. a. $y = 2 + x$ is a model for a relationship of type direct superposition as it has the form of $y = k + x$, $k \in \mathbb{R}$ with $k = 2$.

    b. $y = -6 + x$ is a model for a relationship of type direct superposition as it has the form of $y = k + x$, $k \in \mathbb{R}$ with $k = -6$.

    c. $y = x$ may be rewritten as $y = 0 + x$ and is therefore a model for a relationship of type direct superposition as it has the form of $y = k + x$, $k \in \mathbb{R}$ with $k = 0$.

d. $y = 2 - x$ is not a model for a relationship of type direct superposition as it does not have the form of $y = k + x$, $k \in \mathbb{R}$. Note that in the formal model for direct superposition, x is being added on the right side of the equation whereas in $y = 2 - x$, x is being subtracted on the right side of the equation.

e. $y = \frac{1}{3} + x$ is a model for a relationship of type direct superposition as it has the form of $y = k + x$, $k \in \mathbb{R}$ with $k = \frac{1}{3}$.

f. Rearranging the equation $y + x = 4$ to $y = 4 - x$ shows that it is not a model for a relationship of type direct superposition as it does not have the form of $y = k + x$, $k \in \mathbb{R}$. Note that in the formal model for direct superposition, x is being added on the right side of the equation whereas in $y = 4 - x$, x is being subtracted on the right side of the equation.[13]

g. $y - x = -3$ may be rewritten as $y = -3 + x$ and is therefore a model for a relationship of type direct superposition as it has the form of $y = k + x$, $k \in \mathbb{R}$ with $k = -3$.

h. $y = 4 + 2x$ is not a model for a relationship of type direct superposition as it does not have the form of $y = k + x$, $k \in \mathbb{R}$. Note that in the formal form x appears as a term while in $y = 4 + 2x$ this is not so.[14]

i. $y = -5x$ is not a model for a relationship of type direct superposition as it does not have the form of $y = k + x$, $k \in \mathbb{R}$ even if it is written as $y = 0 - 5x$. Note that in the formal form x appears as a term while in $y = 0 - 5x$ this is not so. Furthermore, in the formal form the term that contains x is being added whereas in $y = 0 - 5x$, the term that contains x is being subtracted.

j. $x - y = 0$ may be rewritten as $y = 0 + x$:

$$
\begin{aligned}
x - y &= 0 \\
-y &= 0 - x \\
y &= 0 + x
\end{aligned}
$$

and is therefore a model for a relationship of type direct superposition as it has the form of $y = k + x$, $k \in \mathbb{R}$ with $k = 0$.[15]

---

[13] We can also compare a given equation to the conservation form of the model for direct superposition. As an example, we can argue that the equation $y + x = 4$ is not a model for relationships of type direct superposition since it does not match the conservation form of the equation for direct superposition, i.e., $y - x = k$, $k \in \mathbb{R}$. Note that in the conservation model for a relationship of type *direct superposition*, i.e., $y - x = k$, $k \in \mathbb{R}$, x is being subtracted on the left side of the equation whereas in $y + x = 4$, x is being added on the left side of the equation.

While it is always possible and sometimes preferable to compare a given equation to the conservation form, for the sake of uniformity we will use the formal form for such comparisons but we do advise the reader to always look for this alternative line of reasoning.

[14] Note that, in $y = 4 + 2x$, an increase in x by a certain amount results in an increase in y, not by the same amount, but *twice* that amount.

[15] We draw the reader's attention to alternative lines of reasoning. Here, it would be easier to appeal to the conservation form of the model for direct superposition to show that the equation $x - y = 0$ is a model for a direct superposition problem. The order of subtraction does not matter as direct superposition is symmetric.

k. $-x + y = \sqrt{3}$ may be rewritten as $y = \sqrt{3} + x$ and is therefore a model for a relationship of type direct superposition as it has the form of $y = k + x$, $k \in \mathbb{R}$ with $k = \sqrt{3}$.

l. $y = x - 1$ may be rewritten as $y = -1 + x$ and is therefore a model for a relationship of type direct superposition as it has the form of $y = k + x$, $k \in \mathbb{R}$ with $k = -1$.

m. $y = -x + 4$ is not a model for a relationship of type direct superposition as it does not have the form of $y = k + x$, $k \in \mathbb{R}$ even if it is written as $y = 4 - x$. Note that in the formal form x is being added on the right side of the equation whereas in $y = -x + 4$ or its rearranged form $y = 4 - x$, x is being subtracted on the right side of the equation.

n. $y = x + \frac{1}{3}$ may be rewritten as $y = \frac{1}{3} + x$ and is therefore a model for a relationship of type direct superposition as it has the form of $y = k + x$, $k \in \mathbb{R}$ with $k = \frac{1}{3}$.

o. $y - 3 = x$ may be rewritten as $y = 3 + x$ and is therefore a model for a relationship of type direct superposition as it has the form of $y = k + x$, $k \in \mathbb{R}$ with $k = 3$.

p. $x = -4 + y$ may be rewritten as $y = 4 + x$:

$$x = -4 + y$$
$$-4 + y = x$$
$$y = 4 + x$$

and is therefore a model for a relationship of type direct superposition as it has the form of $y = k + x$, $k \in \mathbb{R}$ with $k = 4$.

q. $x = 2 - y$ is not a model for a relationship of type direct superposition as it does not have the form of $y = k + x$, $k \in \mathbb{R}$ even if it is written as $y = 2 - x$. Note that in the formal form x is being added on the right side of the equation whereas in $y = 2 - x$, x is being subtracted on the right side of the equation.

r. $-2x = 5 + y$ is not a model for a relationship of type direct superposition as it does not have the form of $y = k + x$, $k \in \mathbb{R}$ even if it is written as $y = -5 - 2x$:

$$-2x = 5 + y$$
$$-y = 5 + 2x$$
$$y = -5 - 2x$$

Note that in the formal form x appears as a term while in $y = -5 - 2x$ this is not so. Furthermore, in the formal form the term that contains x is being added whereas in $y = -5 - 2x$, the term that contains x is being subtracted.

s. $2y = 5 + 2x$ may be rewritten as $y = \frac{5}{2} + x$:

$$2y = 5 + 2x$$
$$y = \frac{1}{2}(5 + 2x)$$
$$y = \frac{5}{2} + x$$

which shows that it is a model for a relationship of type direct superposition as it has the form of $y = k + x$, $k \in \mathbb{R}$ with $k = \frac{5}{2}$.

t. Rearranging the equation $3y = -3 + 4x$ to $y = -1 + \frac{4}{3}x$:

$$3y = -3 + 4x$$

$$y = \frac{1}{3}(-3 + 4x)$$

$$y = -1 + \frac{4}{3}x$$

shows that it is not a model for a relationship of type direct superposition as it does not have the form of $y = k + x$, $k \in \mathbb{R}$. Note that in the formal form x appears as a term while in $y = -1 + \frac{4}{3}x$ this is not so.

12. a. Formal form:

$$y = 7 + x$$

Conservation form:

$$y = 7 + x$$
$$y - x = 7$$

b. Formal form:

$$y = -2 + x$$

Conservation form:

$$y = -2 + x$$
$$y - x = -2$$

c. Formal form:

$$y = x$$
$$y = 0 + x$$

Conservation form:

$$y = x$$
$$y - x = 0$$

d. Formal form:

$$x = 3 + y$$
$$3 + y = x$$
$$y = -3 + x$$

Conservation form:

$$x = 3 + y$$
$$3 + y = x$$
$$y - x = -3$$

e. Formal form:

$$y = x + \frac{2}{5}$$
$$y = \frac{2}{5} + x$$

Conservation form:

$$y = x + \frac{2}{5}$$

$$y - x = \frac{2}{5}$$

f. Formal form:

$$y - x = 4$$
$$y = 4 + x$$

Conservation form:

$$y - x = 4$$

g. Formal form:

$$x - y = -1$$
$$-y = -1 - x$$
$$y = 1 + x$$

Conservation form:

$$x - y = -1$$
$$-x + y = 1$$
$$y - x = 1$$

h. Formal form:

$$y + 4 - x = 0$$
$$y = 0 - 4 + x$$
$$y = -4 + x$$

Conservation form:

$$y + 4 - x = 0$$
$$y - x = 0 - 4$$
$$y - x = -4$$

i. Formal form:

$$y - x - 2 = 0$$
$$y = 0 + x + 2$$
$$y = 2 + x$$

Conservation form:

$$y - x - 2 = 0$$
$$y - x = 0 + 2$$
$$y - x = 2$$

j. Formal form:

$$x - y = 0$$
$$-y = 0 - x$$
$$y = 0 + x$$

Conservation form:

$$x - y = 0$$
$$-x + y = 0$$
$$y - x = 0$$

k. Formal form:

$$-y + x = \sqrt{2}$$
$$-y = \sqrt{2} - x$$
$$y = -\sqrt{2} + x$$

Conservation form:

$$-y + x = \sqrt{2}$$
$$y - x = -\sqrt{2}$$

l. Formal form:

$$-x + 1 = -y$$
$$y = -1 + x$$

Conservation form:

$$-x + 1 = -y$$
$$y - x = -1$$

13. In each solution below, we use our earlier finding that the x- and y-intercepts of $y = k + x$, $k \in \mathbb{R}$ are $(-k, 0)$ and $(0, k)$ to find the x- and y-intercepts of the given line. In addition, to find the slope, we will use the x-intercept as P_1 and the y-intercept as P_2 unless the line goes through the origin in which case the x- and y-intercepts are both $(0, 0)$ and, therefore, we need an additional point. To graph the line we use the x- and y-intercepts unless the line goes through the origin in which case the x- and y-intercepts are both $(0, 0)$ and, therefore, we will need an additional point.

a. Formal form:

$$y = 7 + x$$

x-intercept: $(-7, 0)$
y-intercept: $(0, 7)$

Slope:

$$m = \frac{y_2 - y_1}{x_2 - x_1}$$
$$m = \frac{7 - 0}{0 - (-7)}$$
$$m = \frac{7}{0 + 7}$$
$$m = \frac{7}{7}$$
$$m = 1$$

Graph:

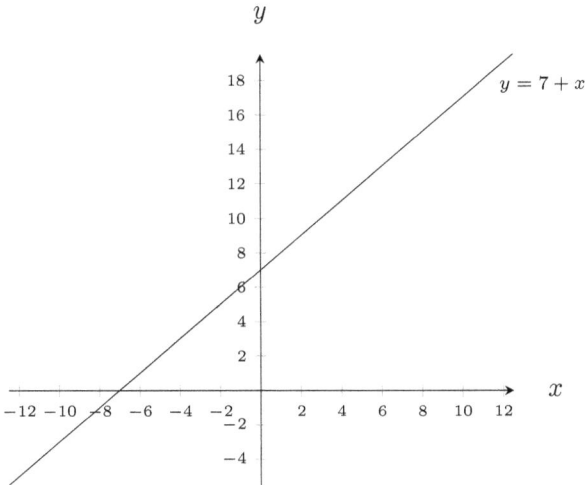

b. Formal form:

$$y = -2 + x$$

x-intercept: $(2, 0)$
y-intercept: $(0, -2)$

Slope:

$$m = \frac{y_2 - y_1}{x_2 - x_1}$$
$$m = \frac{-2 - 0}{0 - 2}$$
$$m = \frac{-2}{-2}$$
$$m = 1$$

Graph:

c. Formal form:

$$y = x$$
$$y = 0 + x$$

x-intercept: $(0, 0)$
y-intercept: $(0, 0)$

Slope: Since the x- and y-intercepts are both $(0, 0)$, we need another point on the line to find the slope of the line. We pick $x = 1$ and use the equation $y = x$ to find y:

$$y = x$$
$$y = 1$$

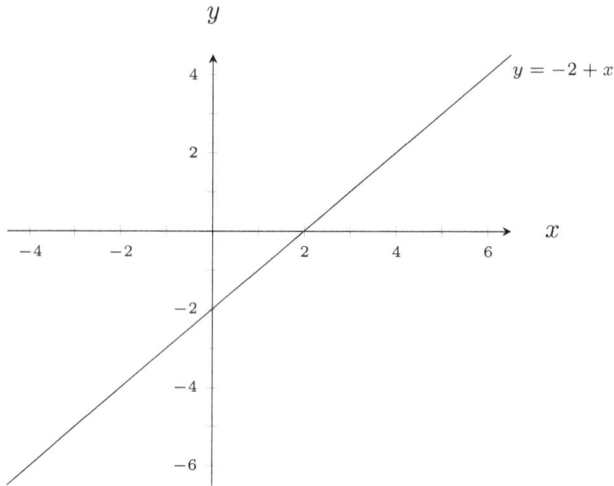

which shows that the point $(1, 1)$ is on the line. We set one of the intercepts as P_1 and the point $(1, 1)$ as P_2. The slope is

$$m = \frac{y_2 - y_1}{x_2 - x_1}$$

$$m = \frac{1 - 0}{1 - 0}$$

$$m = \frac{1}{1}$$

$$m = 1$$

Graph:

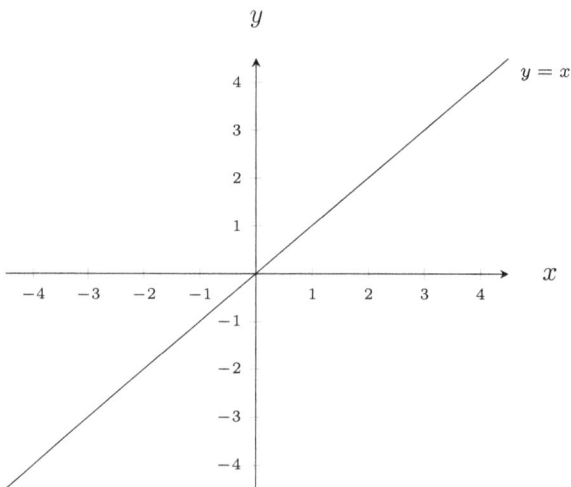

d. Formal form:

$$x = 3 + y$$
$$3 + y = x$$
$$y = -3 + x$$

x-intercept: $(3, 0)$
y-intercept: $(0, -3)$

Slope:

$$m = \frac{y_2 - y_1}{x_2 - x_1}$$
$$m = \frac{-3 - 0}{0 - 3}$$
$$m = \frac{-3}{-3}$$
$$m = 1$$

Graph:

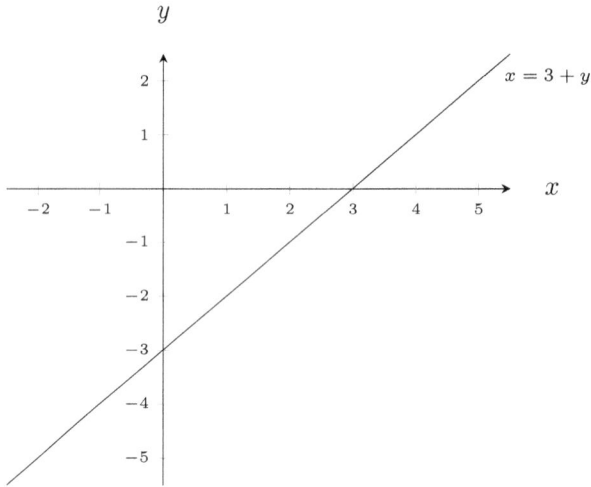

e. Formal form:

$$y = x + \frac{2}{5}$$
$$y = \frac{2}{5} + x$$

x-intercept: $\left(-\frac{2}{5}, 0\right)$

y-intercept: $\left(0, \frac{2}{5}\right)$

Slope:

$$m = \frac{y_2 - y_1}{x_2 - x_1}$$

$$m = \frac{\frac{2}{5} - 0}{0 - \left(-\frac{2}{5}\right)}$$

$$m = \frac{\frac{2}{5}}{0 + \frac{2}{5}}$$

$$m = \frac{\frac{2}{5}}{\frac{2}{5}}$$

$$m = \frac{2 \times 5}{5 \times 2}$$

$$m = 1$$

Graph:

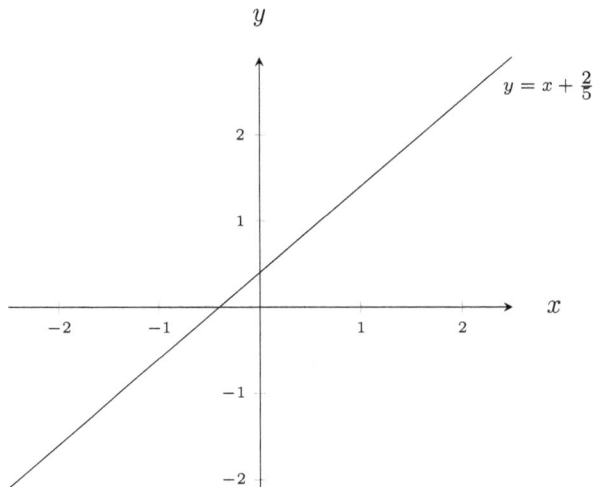

$$y = x + \tfrac{2}{5}$$

f. Formal form:

$$y - x = 4$$
$$y = 4 + x$$

x-intercept: $(-4, 0)$
y-intercept: $(0, 4)$

Slope:

$$m = \frac{y_2 - y_1}{x_2 - x_1}$$
$$m = \frac{4 - 0}{0 - (-4)}$$
$$m = \frac{4}{0 + 4}$$
$$m = \frac{4}{4}$$
$$m = 1$$

Graph:

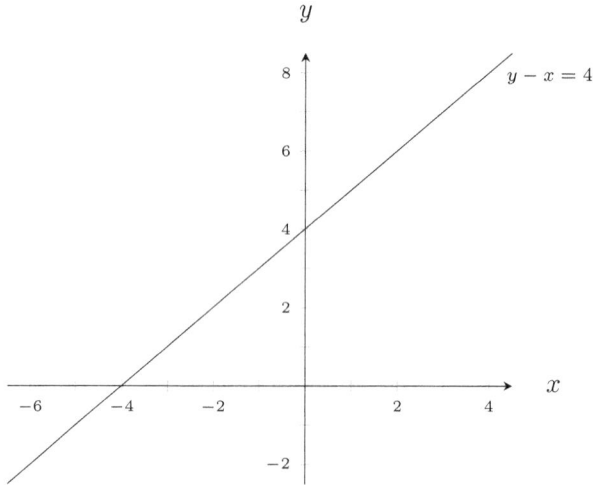

g. Formal form:

$$x - y = -1$$
$$- y = -1 - x$$
$$y = 1 + x$$

x-intercept: $(-1, 0)$
y-intercept: $(0, 1)$

Slope:

$$m = \frac{y_2 - y_1}{x_2 - x_1}$$
$$m = \frac{1 - 0}{0 - (-1)}$$
$$m = \frac{1}{0 + 1}$$
$$m = \frac{1}{1}$$
$$m = 1$$

Graph:

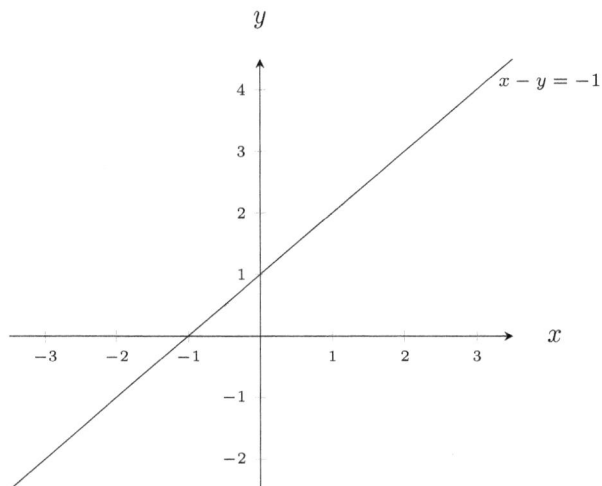

h. Formal form:

$$y + 4 - x = 0$$
$$y = 0 - 4 + x$$
$$y = -4 + x$$

x-intercept: $(4, 0)$
y-intercept: $(0, -4)$

Slope:

$$m = \frac{y_2 - y_1}{x_2 - x_1}$$
$$m = \frac{-4 - 0}{0 - 4}$$
$$m = \frac{-4}{-4}$$
$$m = 1$$

Graph:

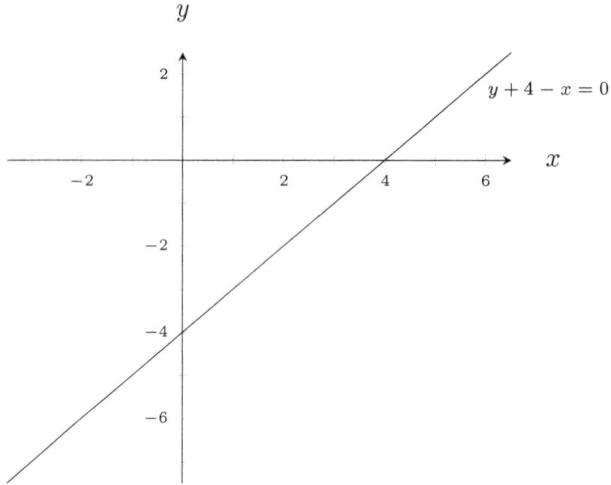

i. Formal form:

$$y - x - 2 = 0$$
$$y = 0 + 2 + x$$
$$y = 2 + x$$

x-intercept: $(-2, 0)$
y-intercept: $(0, 2)$

Slope:

$$m = \frac{y_2 - y_1}{x_2 - x_1}$$
$$m = \frac{2 - 0}{0 - (-2)}$$
$$m = \frac{2}{0 + 2}$$
$$m = \frac{2}{2}$$
$$m = 1$$

Graph:

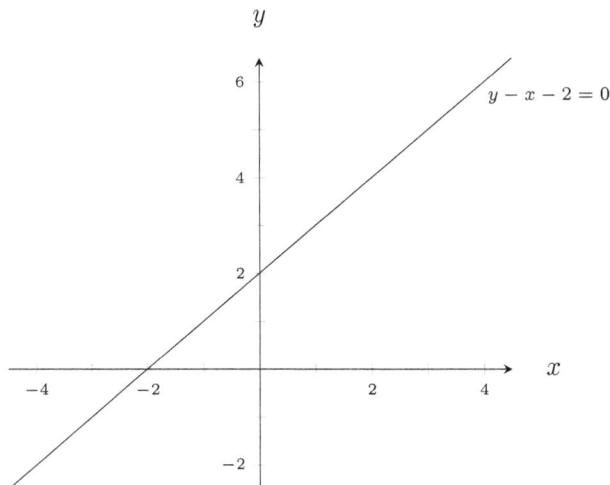

j. Formal form:

$$x - y = 0$$
$$-y = 0 - x$$
$$y = 0 + x$$

x-intercept: $(0,0)$
y-intercept: $(0,0)$

Slope: Since the x- and y-intercepts are both $(0,0)$, we need another point on the line to find the slope of the line. We pick $x = 1$ and use the equation $x - y = 0$ to find y:

$$x - y = 0$$
$$1 - y = 0$$
$$-y = 0 - 1$$
$$-y = -1$$
$$y = 1$$

which shows that the point $(1,1)$ is on the line. We set one of the intercepts as P_1 and the point $(1,1)$ as P_2. The slope is

$$m = \frac{y_2 - y_1}{x_2 - x_1}$$
$$m = \frac{1 - 0}{1 - 0}$$
$$m = \frac{1}{1}$$
$$m = 1$$

Graph:

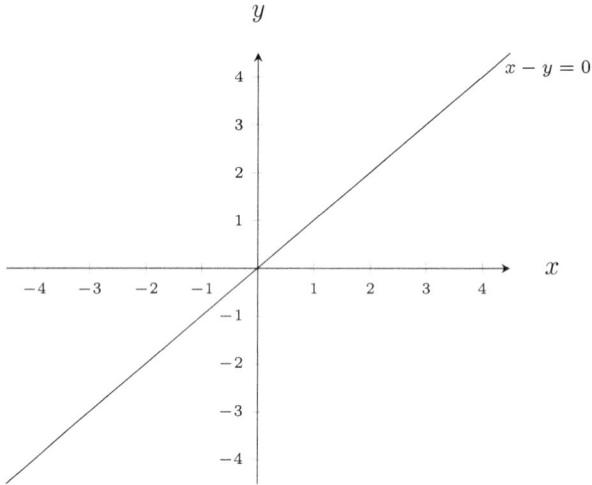

k. Formal form:

$$-y + x = \sqrt{2}$$
$$-y = \sqrt{2} - x$$
$$y = -\sqrt{2} + x$$

x-intercept: $\left(\sqrt{2}, 0\right)$
y-intercept: $\left(0, -\sqrt{2}\right)$

Slope:

$$m = \frac{y_2 - y_1}{x_2 - x_1}$$
$$m = \frac{-\sqrt{2} - 0}{0 - \sqrt{2}}$$
$$m = \frac{-\sqrt{2}}{-\sqrt{2}}$$
$$m = 1$$

Graph:

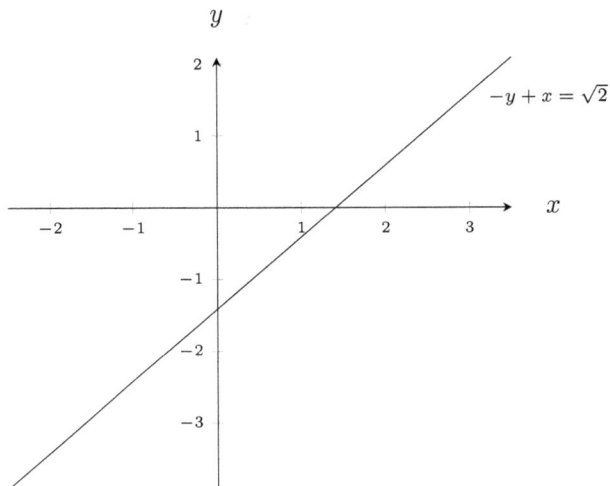

1. Formal form:

$$-x + 1 = -y$$
$$y = -1 + x$$

x-intercept: $(1, 0)$
y-intercept: $(0, -1)$

Slope:

$$m = \frac{y_2 - y_1}{x_2 - x_1}$$
$$m = \frac{-1 - 0}{0 - 1}$$
$$m = \frac{-1}{-1}$$
$$m = 1$$

Graph:

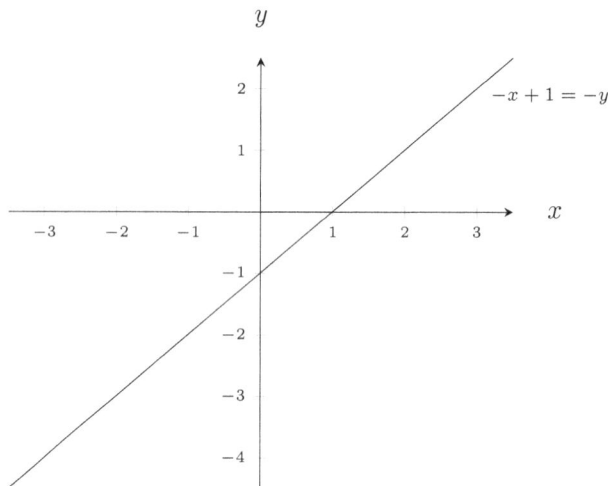

Exercise Set 12.2.1.2

1. It means that an increase or decrease in the value of one of the quantities forces a corresponding decrease or increase in the value of the other quantity.

2. It means that an increase or decrease in the value of one of the quantities by a certain amount forces a corresponding decrease or increase in the value of the other quantity by the same amount.

3. $y = k - x, \quad k \in \mathbb{R}$

4. $y + x = k, \quad k \in \mathbb{R}$

5. It means that an increase or decrease in the value of either quantity by a certain amount forces a corresponding decrease or increase in the value of the other quantity by the same amount.

6. $\Delta y = -\Delta x$ where x and y represent the values of the two quantities.

7. The shape of the graph of the model for a relationship of type inverse super-position is a straight line.

8. -1

9. $-45\,°$

10. The effect of k in $y = k - x$, $k \in \mathbb{R}$, is to raise or lower the line. The higher the value of k, the higher the line.

11. a. $y = 8 - x$ is a model for a relationship of type inverse superposition as it has the form of $y = k - x$, $k \in \mathbb{R}$ with $k = 8$.

    b. $y = -3 - x$ is a model for a relationship of type inverse superposition as it has the form of $y = k - x$, $k \in \mathbb{R}$ with $k = -3$.

c. $y = -x$ may be rewritten as $y = 0 - x$ and is therefore a model for a relationship of type inverse superposition as it has the form of $y = k - x$, $k \in \mathbb{R}$ with $k = 0$.

d. $y = 5 + x$ is not a model for a relationship of type inverse superposition as it does not have the form of $y = k - x$, $k \in \mathbb{R}$. Note that in the formal model for inverse superposition, x is being subtracted on the right side of the equation whereas in $y = 5 + x$, x is being added on the right side of the equation.

e. $y = \frac{2}{5} - x$ is a model for a relationship of type inverse superposition as it has the form of $y = k - x$, $k \in \mathbb{R}$ with $k = \frac{2}{5}$.

f. Rearranging the equation $y - x = 10$ to $y = 10 + x$ shows that it is not a model for a relationship of type inverse superposition as it does not have the form of $y = k - x$, $k \in \mathbb{R}$. Note that in the formal model for inverse superposition, x is being subtracted on the right side of the equation whereas in $y = 10 + x$, x is being added on the right side of the equation.[16]

g. $y + x = -5$ may be rewritten as $y = -5 - x$ and is therefore a model for a relationship of type inverse superposition as it has the form of $y = k - x$, $k \in \mathbb{R}$ with $k = -5$.

h. $y = 6 - 3x$ is not a model for a relationship of type inverse superposition as it does not have the form of $y = k - x$, $k \in \mathbb{R}$. Note that in the formal form x appears as a term while in $y = 6 - 3x$ this is not so.[17]

i. $y = 9x$ is not a model for a relationship of type inverse superposition as it does not have the form of $y = k - x$, $k \in \mathbb{R}$ even if it is written as $y = 0 + 9x$. Note that in the formal form x appears as a term while in $y = 0 + 9x$ this is not so. Furthermore, in the formal form the term that contains x is being subtracted whereas in $y = 0 + 9x$, the term that contains x is being added.

j. $x + y = 0$ may be rewritten as $y = 0 - x$ and is therefore a model for a relationship of type inverse superposition as it has the form of $y = k - x$, $k \in \mathbb{R}$ with $k = 0$.[18]

---

[16] We can also compare a given equation to the conservation form of the model for inverse superposition. As an example, we can argue that the equation $y - x = 10$ is not a model for relationships of type inverse superposition since it does not match the conservation form of the equation for inverse superposition, i.e., $y + x = k$, $k \in \mathbb{R}$. Note that in the conservation model for a relationship of type *inverse superposition*, i.e., $y + x = k$, $k \in \mathbb{R}$, x is being added on the left side of the equation whereas in $y - x = 10$, x is being subtracted on the left side of the equation.

While it is always possible and sometimes preferable to compare a given equation to the conservation form, for the sake of uniformity we will use the formal form for such comparisons but we do advise the reader to always look for this alternative line of reasoning.

[17] Note that an increase in x by a certain amount results in a decrease in y, not by the same amount, but *three times* that amount.

[18] We draw the reader's attention to alternative lines of reasoning. Here, it would be easier to appeal to the conservation form of the model for inverse superposition to show that the equation $x + y = 0$ is a model for an inverse superposition problem. The order of addition does not matter as inverse superposition is symmetric.

k. $-x - y = \sqrt{2}$ may be rewritten as $y = -\sqrt{2} - x$:

$$
\begin{aligned}
- x - y &= \sqrt{2} \\
- y &= \sqrt{2} + x \\
y &= -\sqrt{2} - x
\end{aligned}
$$

and is therefore a model for a relationship of type inverse superposition as it has the form of $y = k - x$, $k \in \mathbb{R}$ with $k = -\sqrt{2}$.

l. $y = -x - 1$ may be rewritten as $y = -1 - x$ and is therefore a model for a relationship of type inverse superposition as it has the form of $y = k - x$, $k \in \mathbb{R}$ with $k = -1$.

m. $y = x + 5$ is not a model for a relationship of type inverse superposition as it does not have the form of $y = k - x$, $k \in \mathbb{R}$ even if it is written as $y = 5 + x$. Note that in the formal form x is being subtracted on the right side of the equation whereas in $y = x + 5$ or its rearranged form $y = 5 + x$, x is being added on the right side of the equation.

n. $y = -x - \frac{3}{4}$ may be rewritten as $y = -\frac{3}{4} - x$ and is therefore a model for a relationship of type inverse superposition as it has the form of $y = k - x$, $k \in \mathbb{R}$ with $k = -\frac{3}{4}$.

o. $y - 3 = -x$ may be rewritten as $y = 3 - x$ and is therefore a model for a relationship of type inverse superposition as it has the form of $y = k - x$, $k \in \mathbb{R}$ with $k = 3$.

p. $x = -7 - y$ may be rewritten as $y = -7 - x$ and is therefore a model for a relationship of type inverse superposition as it has the form of $y = k - x$, $k \in \mathbb{R}$ with $k = -7$.

q. $-x = 1 - y$ is not a model for a relationship of type inverse superposition as it does not have the form of $y = k - x$, $k \in \mathbb{R}$ even if it is written as $y = 1 + x$. Note that in the formal form x is being subtracted on the right side of the equation whereas in $y = 1 + x$, x is being added on the right side of the equation.

r. $-3x = 8 - y$ is not a model for a relationship of type inverse superposition as it does not have the form of $y = k - x$, $k \in \mathbb{R}$ even if it is written as $y = 8 + 3x$. Note that in the formal form x appears as a term while in $y = 8 + 3x$ this is not so. Furthermore, in the formal form the term that contains x is being subtracted whereas in $y = 8 + 3x$, the term that contains x is being added.

s. $-5y = -1 + 5x$ may be rewritten as $y = \frac{1}{5} - x$:

$$
\begin{aligned}
- 5y &= -1 + 5x \\
y &= -\frac{1}{5}(-1 + 5x) \\
y &= \frac{1}{5} - x
\end{aligned}
$$

which shows that it is a model for a relationship of type inverse superposition as it has the form of $y = k - x$, $k \in \mathbb{R}$ with $k = \frac{1}{5}$.

t. Rearranging the equation $-2y = -3 - 2x$ to $y = \frac{3}{2} + x$:

$$-2y = -3 - 2x$$

$$y = -\frac{1}{2}(-3 - 2x)$$

$$y = \frac{3}{2} + x$$

shows that it is not a model for a relationship of type inverse superposition as it does not have the form of $y = k - x$, $k \in \mathbb{R}$. Note that in the formal form x is subtracted on the right side of the equation whereas in $y = \frac{3}{2} + x$, x is being added on the right side of the equation.

12. a. Formal form:

$$y = 7 - x$$

Conservation form:

$$y = 7 - x$$
$$y + x = 7$$

b. Formal form:

$$y = -2 - x$$

Conservation form:

$$y = -2 - x$$
$$y + x = -2$$

c. Formal form:

$$y = -x$$
$$y = 0 - x$$

Conservation form:

$$y = -x$$
$$y + x = 0$$

d. Formal form:

$$x = 3 - y$$
$$y = 3 - x$$

Conservation form:

$$x = 3 - y$$
$$y + x = 3$$

e. Formal form:

$$y = -x + \frac{2}{5}$$
$$y = \frac{2}{5} - x$$

Conservation form:

$$y = -x + \frac{2}{5}$$
$$y + x = \frac{2}{5}$$

f. Formal form:

$$y + x = 4$$
$$y = 4 - x$$

Conservation form:

$$y + x = 4$$

g. Formal form:

$$x + y = -1$$
$$y = -1 - x$$

Conservation form:

$$x + y = -1$$
$$y + x = -1$$

h. Formal form:

$$y + 4 + x = 0$$
$$y = 0 - 4 - x$$
$$y = -4 - x$$

Conservation form:

$$y + 4 + x = 0$$
$$y + x = 0 - 4$$
$$y + x = -4$$

i. Formal form:

$$y + x - 2 = 0$$
$$y = 0 - x + 2$$
$$y = 2 - x$$

Conservation form:

$$y + x - 2 = 0$$
$$y + x = 0 + 2$$
$$y + x = 2$$

j. Formal form:

$$x + y = 0$$
$$y = 0 - x$$

Conservation form:

$$x + y = 0$$
$$y + x = 0$$

k. Formal form:

$$-y - x = \sqrt{2}$$
$$-y = \sqrt{2} + x$$
$$y = -\sqrt{2} - x$$

Conservation form:

$$-y - x = \sqrt{2}$$
$$y + x = -\sqrt{2}$$

l. Formal form:

$$x + 1 = -y$$
$$y = -1 - x$$

Conservation form:

$$x + 1 = -y$$
$$y + x = -1$$

13. In each solution below, we use our earlier finding that the x- and y-intercepts of $y = k - x$, $k \in \mathbb{R}$ are $(k, 0)$ and $(0, k)$ to find the x- and y-intercepts of the given line. In addition, to find the slope, we will use the x-intercept as P_1 and the y-intercept as P_2 unless the line goes through the origin in which case the x- and y-intercepts are both $(0, 0)$ and, therefore, we need an additional point. To graph the line we use the x- and y-intercepts unless the line goes through the origin in which case the x- and y-intercepts are both $(0, 0)$ and, therefore, we will need an additional point.

a. Formal form:

$$y = 7 - x$$

x-intercept: $(7, 0)$
y-intercept: $(0, 7)$

Slope:

$$m = \frac{y_2 - y_1}{x_2 - x_1}$$
$$m = \frac{7 - 0}{0 - 7}$$
$$m = \frac{7}{-7}$$
$$m = -1$$

Graph:

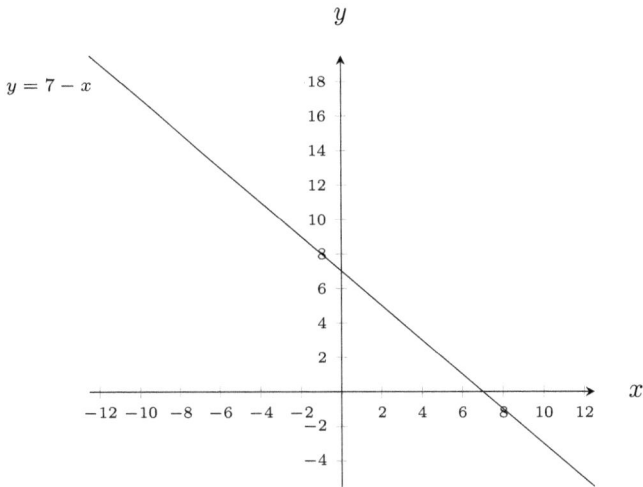

$y = 7 - x$

b. Formal form:

$$y = -2 - x$$

x-intercept: $(-2, 0)$
y-intercept: $(0, -2)$

Slope:

$$m = \frac{y_2 - y_1}{x_2 - x_1}$$

$$m = \frac{-2 - 0}{0 - (-2)}$$

$$m = \frac{-2}{0 + 2}$$

$$m = \frac{-2}{2}$$

$$m = -1$$

Graph:

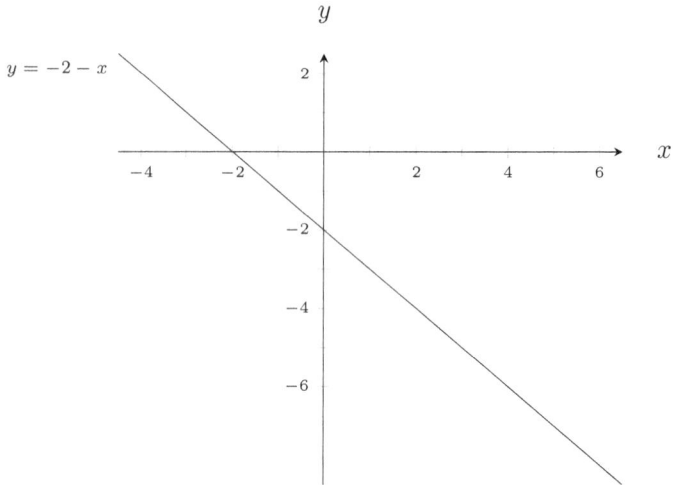

c. Formal form:

$$y = -x$$
$$y = 0 - x$$

x-intercept: $(0, 0)$
y-intercept: $(0, 0)$

Slope: Since the x- and y-intercepts are both $(0, 0)$, we need another point on the line to find the slope of the line. We pick $x = 1$ and use the equation $y = -x$ to find y:

$$y = -x$$
$$y = -1$$

which shows that the point $(1, -1)$ is on the line. We set one of the intercepts as P_1 and the point $(1, -1)$ as P_2. The slope is

$$m = \frac{y_2 - y_1}{x_2 - x_1}$$
$$m = \frac{-1 - 0}{1 - 0}$$
$$m = \frac{-1}{1}$$
$$m = -1$$

Graph:

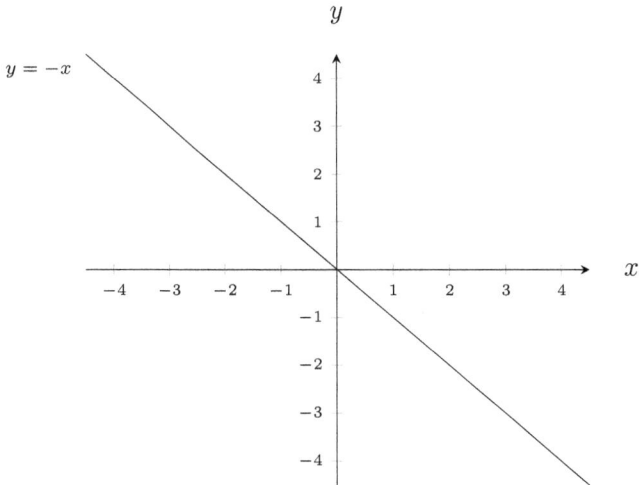

$y = -x$

d. Formal form:

$$x = 3 - y$$
$$y = 3 - x$$

x-intercept: $(3, 0)$
y-intercept: $(0, 3)$

Slope:

$$m = \frac{y_2 - y_1}{x_2 - x_1}$$
$$m = \frac{3 - 0}{0 - 3}$$
$$m = \frac{3}{-3}$$
$$m = -1$$

Graph:

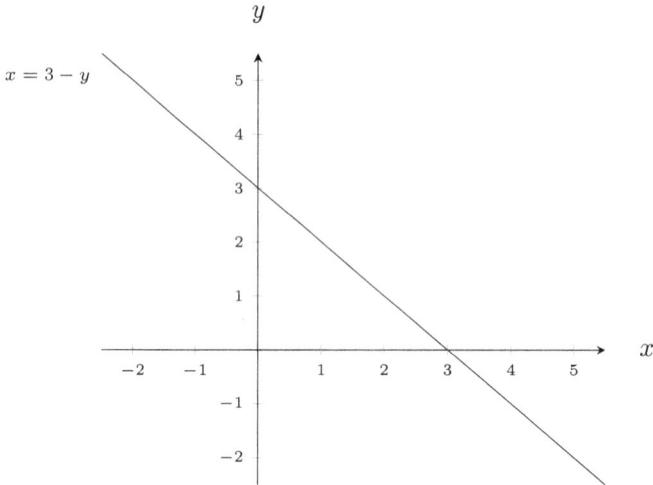

e. Formal form:

$$y = -x + \frac{2}{5}$$
$$y = \frac{2}{5} - x$$

x-intercept: $\left(\frac{2}{5}, 0\right)$

y-intercept: $\left(0, \frac{2}{5}\right)$

Slope:

$$m = \frac{y_2 - y_1}{x_2 - x_1}$$
$$m = \frac{\frac{2}{5} - 0}{0 - \frac{2}{5}}$$
$$m = \frac{\frac{2}{5}}{-\frac{2}{5}}$$
$$m = -\frac{2 \times 5}{5 \times 2}$$
$$m = -1$$

Graph:

$$y = -x + \tfrac{2}{5}$$

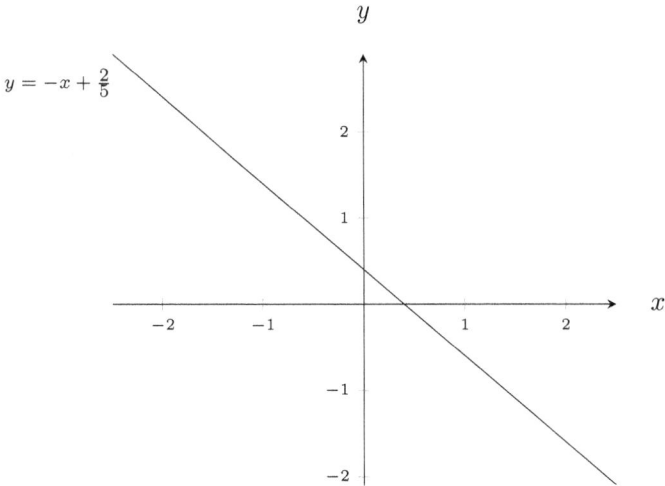

f. Formal form:

$$y + x = 4$$
$$y = 4 - x$$

x-intercept: $(4, 0)$
y-intercept: $(0, 4)$

Slope:

$$m = \frac{y_2 - y_1}{x_2 - x_1}$$
$$m = \frac{4 - 0}{0 - 4}$$
$$m = \frac{4}{-4}$$
$$m = -1$$

Graph:

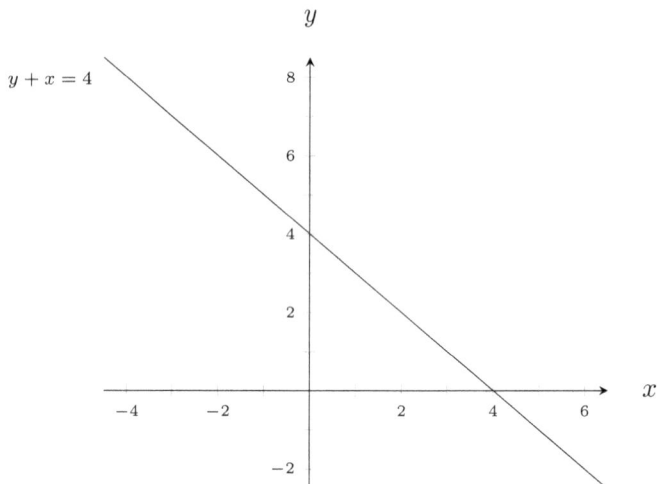

g. Formal form:

$$x + y = -1$$
$$y = -1 - x$$

x-intercept: $(-1, 0)$
y-intercept: $(0, -1)$

Slope:

$$m = \frac{y_2 - y_1}{x_2 - x_1}$$
$$m = \frac{-1 - 0}{0 - (-1)}$$
$$m = \frac{-1}{0 + 1}$$
$$m = \frac{-1}{1}$$
$$m = -1$$

Graph:

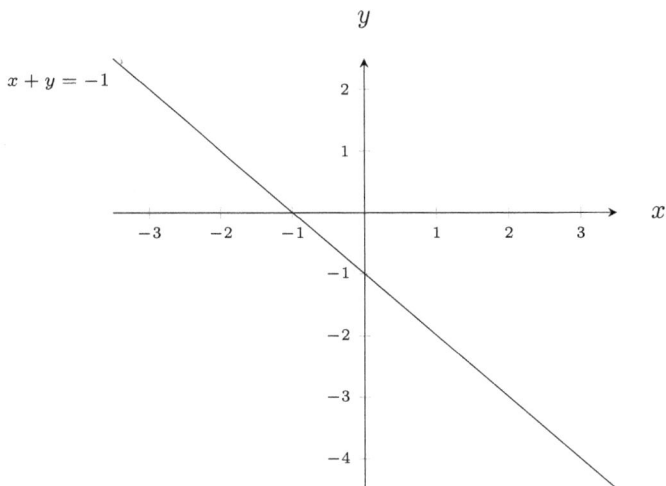

$x + y = -1$

h. Formal form:

$$y + 4 + x = 0$$
$$y = 0 - 4 - x$$
$$y = -4 - x$$

x-intercept: $(-4, 0)$
y-intercept: $(0, -4)$

Slope:

$$m = \frac{y_2 - y_1}{x_2 - x_1}$$
$$m = \frac{-4 - 0}{0 - (-4)}$$
$$m = \frac{-4}{0 + 4}$$
$$m = \frac{-4}{4}$$
$$m = -1$$

Graph:

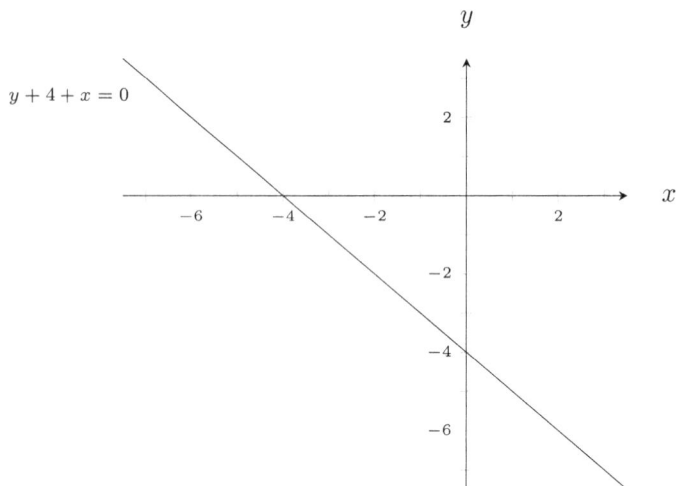

i. Formal form:

$$y + x - 2 = 0$$
$$y = 0 + 2 - x$$
$$y = 2 - x$$

x-intercept: $(2, 0)$
y-intercept: $(0, 2)$

Slope:

$$m = \frac{y_2 - y_1}{x_2 - x_1}$$
$$m = \frac{2 - 0}{0 - 2}$$
$$m = \frac{2}{-2}$$
$$m = -1$$

Graph:

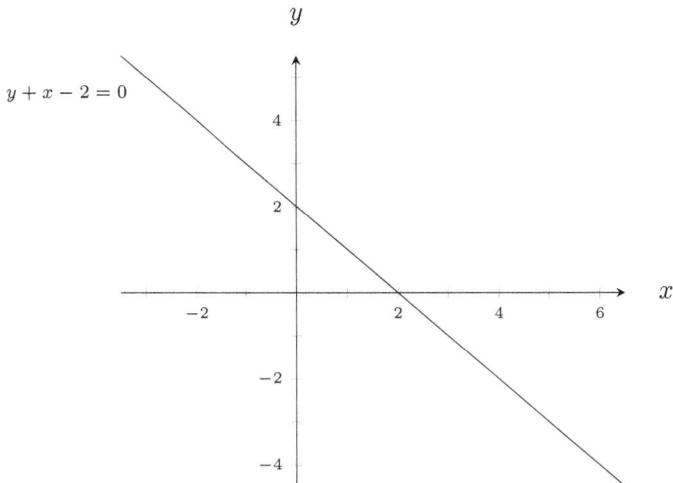

j. Formal form:

$$x + y = 0$$
$$y = 0 - x$$

x-intercept: $(0,0)$
y-intercept: $(0,0)$

Slope: Since the x- and y-intercepts are both $(0,0)$, we need another point on the line to find the slope of the line. We pick $x = 1$ and use the equation $x + y = 0$ to find y:

$$x + y = 0$$
$$1 + y = 0$$
$$y = 0 - 1$$
$$y = -1$$

which shows that the point $(1,-1)$ is on the line. We set one of the intercepts as P_1 and the point $(1,-1)$ as P_2. The slope is

$$m = \frac{y_2 - y_1}{x_2 - x_1}$$
$$m = \frac{-1 - 0}{1 - 0}$$
$$m = \frac{-1}{1}$$
$$m = -1$$

Graph:

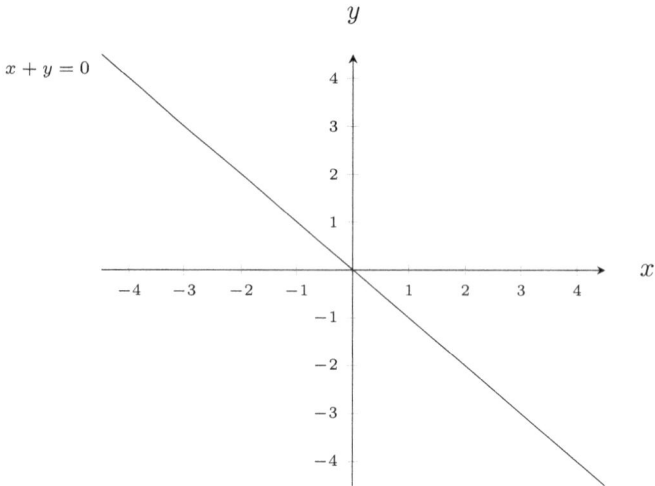

$x + y = 0$

k. Formal form:

$$
\begin{aligned}
-y - x &= \sqrt{2} \\
-y &= \sqrt{2} + x \\
y &= -\sqrt{2} - x
\end{aligned}
$$

x-intercept: $\left(-\sqrt{2}, 0\right)$
y-intercept: $\left(0, -\sqrt{2}\right)$

Slope:

$$
\begin{aligned}
m &= \frac{y_2 - y_1}{x_2 - x_1} \\
m &= \frac{-\sqrt{2} - 0}{0 - \left(-\sqrt{2}\right)} \\
m &= \frac{-\sqrt{2} - 0}{0 + \sqrt{2}} \\
m &= \frac{-\sqrt{2}}{\sqrt{2}} \\
m &= -1
\end{aligned}
$$

Graph:

$-y - x = \sqrt{2}$

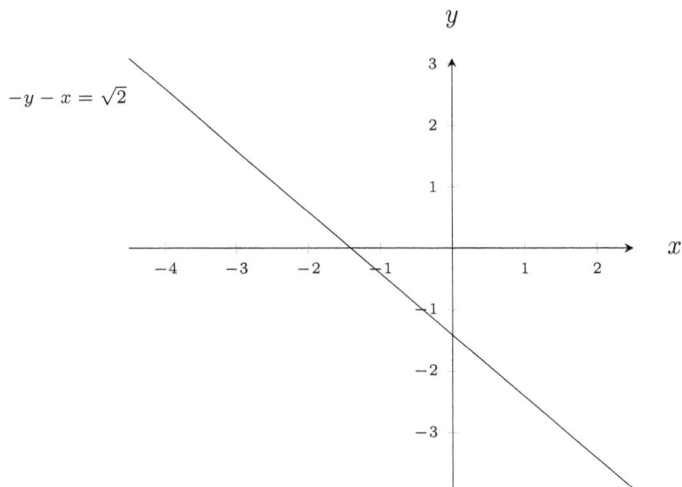

1. Formal form:

$$x + 1 = -y$$
$$y = -1 - x$$

x-intercept: $(-1, 0)$
y-intercept: $(0, -1)$

Slope:

$$m = \frac{y_2 - y_1}{x_2 - x_1}$$
$$m = \frac{-1 - 0}{0 - (-1)}$$
$$m = \frac{-1}{0 + 1}$$
$$m = \frac{-1}{1}$$
$$m = -1$$

Graph:

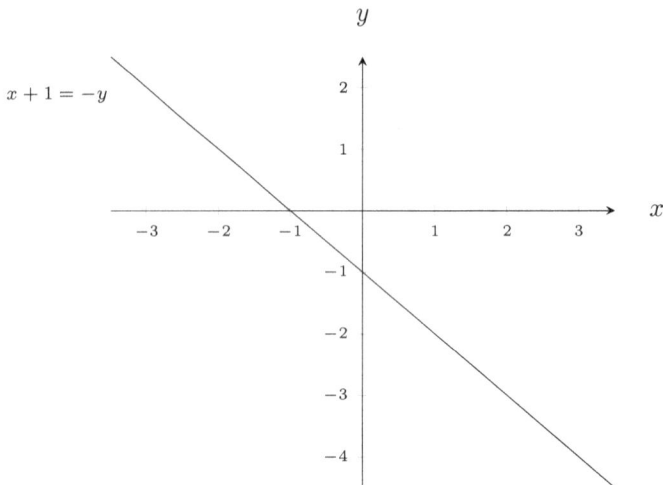

$x + 1 = -y$

Exercise Set 12.2.1.3

1. It means that scaling the value of one of the quantities forces a corresponding scaling of the value of the other quantity with both quantities increasing in size or both quantities decreasing in size.

2. It means that scaling the value of one of the quantities by a certain factor forces a corresponding scaling of the value of the other quantity by the same factor.

3. $y = kx, \quad k \in \mathbb{R}, k \neq 0$

4. $\frac{y}{x} = k, \quad k \in \mathbb{R}, k \neq 0$

5. It means that scaling the value of either quantity by a certain factor forces a scaling of the value of the other quantity by the same factor.

6. $\Delta y = k\Delta x$ where x and y represent the values of the two quantities and k is the constant of proportionality.

7. The shape of the graph of the model for a relationship of type direct proportion is a straight line.

8. k, where k is the constant of proportionality.

9. For $0 < k < 1$, the angle between the line and the horizontal is between $0°$ and $45°$. When $k = 1$, the angle between the line and the horizontal becomes $45°$. For $k > 1$ the angle between the line and the horizontal is greater than $45°$.

10. The effect of k in $y = kx$, $k \in \mathbb{R}$, $k < 0$ is to rotate the line about the origin. The larger the size of k, the steeper the line. The following figure shows the graph of the lines $y = -\frac{1}{2}x$, $y = -x$ and $y = -2x$.

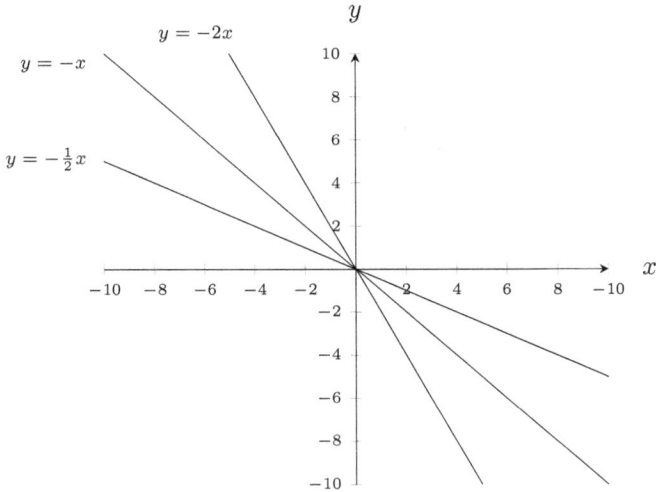

11. a. $y = 2x$ is a model for a relationship of type direct proportion as it has the form of $y = kx$, $k \in \mathbb{R}$, $k \neq 0$ with $k = 2$.

b. $y = -6x$ is a model for a relationship of type direct proportion as it has the form of $y = kx$, $k \in \mathbb{R}$, $k \neq 0$ with $k = -6$.

c. $y = x$ may be rewritten as $y = 1x$ and is therefore a model for a relationship of type direct proportion as it has the form of $y = kx$, $k \in \mathbb{R}$, $k \neq 0$ with $k = 1$.

d. $y = \frac{2}{5}x$ is a model for a relationship of type direct proportion as it has the form of $y = kx$, $k \in \mathbb{R}$, $k \neq 0$ with $k = \frac{2}{5}$.

e. $y = -\frac{1}{3}x$ is a model for a relationship of type direct proportion as it has the form of $y = kx$, $k \in \mathbb{R}$, $k \neq 0$ with $k = -\frac{1}{3}$.

f. Rearranging the equation $\frac{y}{x} = 7$ to $y = 7x$ shows that it is a model for a relationship of type direct proportion as it has the form of $y = kx$, $k \in \mathbb{R}$, $k \neq 0$ with $k = 7$.[19]

g. $\frac{y}{x} = -\frac{3}{4}$ may be rewritten as $y = -\frac{3}{4}x$ and is therefore a model for a relationship of type direct proportion as it has the form of $y = kx$, $k \in \mathbb{R}$, $k \neq 0$ with $k = -\frac{3}{4}$.

h. $yx = -3$ is not a model for a relationship of type direct proportion as it does not have the form of $y = kx$, $k \in \mathbb{R}$, $k \neq 0$ even if it is written as $y = \frac{-3}{x}$. Note that in the formal form x is being multiplied by on the right

---

[19] We can also compare equations to the conservation form of the model for direct proportion. As an example, we can argue that the equation $\frac{y}{x} = 7$ is a model for relationships of type direct proportion since it matches the conservation form of the equation for direct proportion, i.e., $\frac{y}{x} = k$, $k \in \mathbb{R}$, $k \neq 0$.

While it is always possible to compare a given equation to the conservation form, for the sake of uniformity we will use the formal form for such comparisons but we do advise the reader to always look for this alternative line of reasoning.

side of the equation whereas in $y = \frac{-3}{x}$, x is being divided by on the right side of the equation.

i. $xy = \frac{1}{5}$ is not a model for a relationship of type direct proportion as it does not have the form of $y = kx$, $k \in \mathbb{R}$, $k \neq 0$ even if it is written as $y = \frac{1}{5x}$. Note that in the formal form x is being multiplied by on the right side of the equation whereas in $y = \frac{1}{5x}$, x is being divided by on the right side of the equation.[20]

j. $\frac{x}{y} = -1$ may be rewritten as $y = -x$:

$$\frac{x}{y} = -1$$

$$y = \frac{x}{-1}$$

$$y = -x$$

This may be seen as $y = -1x$ and is therefore a model for a relationship of type direct proportion as it has the form of $y = kx$, $k \in \mathbb{R}$, $k \neq 0$ with $k = -1$.[21]

k. $\frac{x}{y} = 4$ may be rewritten as $y = \frac{1}{4}x$:

$$\frac{x}{y} = 4$$

$$y = \frac{x}{4}$$

$$y = \frac{1}{4}x$$

and is therefore a model for a relationship of type direct proportion as it has the form of $y = kx$, $k \in \mathbb{R}$, $k \neq 0$ with $k = \frac{1}{4}$.

l. $y = \frac{x}{8}$ may be rewritten as[22] $y = \frac{1}{8}x$ and is therefore a model for a relationship of type direct proportion as it has the form of $y = kx$, $k \in \mathbb{R}$, $k \neq 0$ with $k = \frac{1}{8}$.

m. $y = -\frac{x}{2}$ may be rewritten as $y = -\frac{1}{2}x$ and is therefore a model for a relationship of type direct proportion as it has the form of $y = kx$, $k \in \mathbb{R}$, $k \neq 0$ with $k = -\frac{1}{2}$.

n. $y = \frac{3}{x}$ is not a model for a relationship of type direct proportion as it does not have the form of $y = kx$, $k \in \mathbb{R}$, $k \neq 0$. Note that in the formal form x is being multiplied by on the right side of the equation whereas in $y = \frac{3}{x}$, x is being divided by on the right side of the equation.

---

[20] We have shown earlier that an expression such as $\frac{1}{5x}$ may be seen as $1 \div 5 \div x$ which shows that x is being divided by.

[21] We draw the reader's attention to alternative lines of reasoning. Here, it would be easier to appeal to the conservation form of the model for direct proportion to show that the equation $\frac{x}{y} = -1$ is a model for a direct proportion problem. The order of division does not matter as direct proportion is symmetric.

[22] We can use semantics to show this equivalence: Note that dividing x into 8 parts (i.e., $\frac{x}{8}$) is the same as finding one-eighth of x (i.e., $\frac{1}{8}x$).

o. $y = -\frac{1}{x}$, which may be seen as $y = \frac{-1}{x}$, is not a model for a relationship of type direct proportion as it does not have the form of $y = kx$, $k \in \mathbb{R}$, $k \neq 0$. Note that in the formal form x is being multiplied by on the right side of the equation whereas in $y = \frac{-1}{x}$, x is being divided by on the right side of the equation.

p. $x = -4y$ maybe written as $y = -\frac{1}{4}x$ and is, therefore, a model for a relationship of type direct proportion as it has the form of $y = kx$, $k \in \mathbb{R}$, $k \neq 0$ with $k = -\frac{1}{4}$.

q. $x = \frac{1}{6}y$ may be rewritten as $y = 6x$ and is therefore a model for a relationship of type direct proportion as it has the form of $y = kx$, $k \in \mathbb{R}$, $k \neq 0$ with $k = 6$.

r. $-2x = 5y$ may be rewritten as $y = -\frac{2}{5}x$ and is therefore a model for a relationship of type direct proportion as it has the form of $y = kx$, $k \in \mathbb{R}$, $k \neq 0$ with $k = -\frac{2}{5}$.

s. $3y = 6x$ may be rewritten as $y = 2x$:

$$3y = 6x$$
$$y = \frac{6}{3}x$$
$$y = 2x$$

and is, therefore, a model for a relationship of type direct proportion as it has the form of $y = kx$, $k \in \mathbb{R}$, $k \neq 0$ with $k = 2$.

t. Rearranging the equation $2y = -8 + 5x$ to $y = -4 + \frac{5}{2}x$:

$$2y = -8 + 5x$$
$$y = \frac{1}{2}(-8 + 5x)$$
$$y = -4 + \frac{5}{2}x$$

shows that it is not a model for a relationship of type direct proportion as it does not have the form of $y = kx$, $k \in \mathbb{R}$, $k \neq 0$. Note that in the formal form the expression on the right side of the equation analyzes into one term whereas the expression on the right side of the equation $y = -4 + \frac{5}{2}x$ analyzes into two terms.

u. Rewriting the equation $3x - 1 = y$ to $y = 3x - 1$ shows that it is not a model for a relationship of type direct proportion as it does not have the form of $y = kx$, $k \in \mathbb{R}$, $k \neq 0$. Note that in the formal form the expression on the right side of the equation analyzes into one term whereas the expression on the right side of the equation $y = 3x - 1$ analyzes into two terms.

v. We can rewrite the equation $-2x - 3y = 0$ as $y = -\frac{2}{3}x$:

$$-2x - 3y = 0$$
$$-3y = 0 + 2x$$
$$-3y = 2x$$
$$y = -\frac{2}{3}x$$

This has the form of $y = kx$, $k \in \mathbb{R}$, $k \neq 0$ with $k = -\frac{2}{3}$ and is, therefore, a model for a relationship of type direct proportion.

w. Rearranging the equation $7x + 2y = 4$ to $y = 2 - \frac{7}{2}x$:

$$7x + 2y = 4$$
$$2y = 4 - 7x$$
$$y = \frac{1}{2}(4 - 7x)$$
$$y = 2 - \frac{7}{2}x$$

shows that it is not a model for a relationship of type direct proportion as it does not have the form of $y = kx$, $k \in \mathbb{R}$, $k \neq 0$. Note that in the formal form the expression on the right side of the equation analyzes into one term whereas the expression on the right side of the equation $y = 2 - \frac{7}{2}x$ analyzes into two terms.

12. a. Formal form:

$$y = 7x$$

Conservation form:

$$y = 7x$$
$$\frac{y}{x} = 7$$

b. Formal form:

$$y = -2x$$

Conservation form:

$$y = -2x$$
$$\frac{y}{x} = -2$$

c. Formal form:

$$y = x$$
$$y = 1x$$

Conservation form:

$$y = x$$
$$\frac{y}{x} = 1$$

d. Formal form:

$$x = 3y$$
$$3y = x$$
$$y = \frac{1}{3}x$$

Conservation form:

$$x = 3y$$
$$3y = x$$
$$\frac{y}{x} = \frac{1}{3}$$

e. Formal form:

$$y = \frac{2}{5}x$$

Conservation form:

$$y = \frac{2}{5}x$$
$$\frac{y}{x} = \frac{2}{5}$$

f. Formal form:

$$\frac{y}{x} = 4$$
$$y = 4x$$

Conservation form:

$$\frac{y}{x} = 4$$

g. Formal form:

$$\frac{x}{y} = -1$$
$$y = \frac{x}{-1}$$
$$y = -x$$
$$y = -1x$$

Conservation form:

$$\frac{x}{y} = -1$$

$$\frac{y}{x} = \frac{1}{-1}$$

$$\frac{y}{x} = -1$$

h. Formal form:

$$4y - x = 0$$
$$4y = 0 + x$$
$$4y = x$$
$$y = \frac{1}{4}x$$

Conservation form:

$$4y - x = 0$$
$$4y = 0 + x$$
$$4y = x$$
$$\frac{y}{x} = \frac{1}{4}$$

i. Formal form:

$$y - x = 0$$
$$y = 0 + x$$
$$y = x$$
$$y = 1x$$

Conservation form:

$$y - x = 0$$
$$y = 0 + x$$
$$y = x$$
$$\frac{y}{x} = 1$$

j. Formal form:

$$x - y = 0$$
$$-y = 0 - x$$
$$-y = -x$$
$$y = x$$
$$y = 1x$$

Conservation form:

$$x - y = 0$$
$$-y = -x$$
$$y = x$$
$$\frac{y}{x} = 1$$

k. Formal form:

$$-\frac{y}{x} = \sqrt{2}$$
$$y = -\sqrt{2}x$$

Conservation form:

$$-\frac{y}{x} = \sqrt{2}$$
$$\frac{y}{x} = -\sqrt{2}$$

l. Formal form:

$$x = -y$$
$$y = -x$$
$$y = -1x$$

Conservation form:

$$x = -y$$
$$y = -x$$
$$\frac{y}{x} = -1$$

13. In each solution below, we use our earlier finding that the x- and y-intercepts of $y = kx$, $k \in \mathbb{R}$, $k \neq 0$ are both $(0,0)$ to find the x- and y-intercepts of the given line. In addition, to find the slope, we can use our earlier findings that, the slope of the line that represents the graph of $y = kx$, $k \in \mathbb{R}$, $k \neq 0$ is k. Here we will use the formula $m = \frac{y_2 - y_1}{x_2 - x_1}$ to find the x- and y-intercepts with the intercept $(0,0)$ as P_1 and find an additional point on the line as P_2. To graph the line we use the x- and y-intercept as one point on the line and find

an additional point on the line as the second point.

a. Formal form:

$$y = 7x$$

x-intercept: $(0, 0)$
y-intercept: $(0, 0)$

Slope: We choose the x- and y-intercept, i.e., $(0, 0)$, as P_1. For P_2, we set $x = 1$ and find the corresponding y value using the equation $y = 7x$:

$$y = 7x$$
$$y = 7 \times 1$$
$$y = 7$$

This shows that the point $(1, 7)$ is on the line. We choose this as P_2. The slope of the line is

$$m = \frac{y_2 - y_1}{x_2 - x_1}$$
$$m = \frac{7 - 0}{1 - 0}$$
$$m = \frac{7}{1}$$
$$m = 7$$

Graph: To graph the line, we choose the x- and y-intercept, i.e., $(0, 0)$, as one point and the point $(1, 7)$ found earlier as our second point.

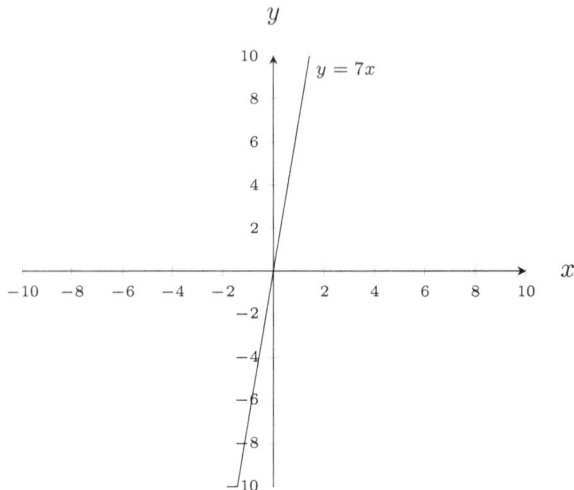

b. Formal form:

$$y = -2x$$

x-intercept: $(0,0)$
y-intercept: $(0,0)$

Slope: We choose the x- and y-intercept, i.e., $(0,0)$, as P_1. For P_2, we set $x = 1$ and find the corresponding y value using the equation $y = -2x$:[23]

$$y = -2x$$
$$y = -2 \times 1$$
$$y = -2$$

This shows that the point $(1, -2)$ is on the line. We choose this as P_2. The slope of the line is

$$m = \frac{y_2 - y_1}{x_2 - x_1}$$
$$m = \frac{-2 - 0}{1 - 0}$$
$$m = \frac{-2}{1}$$
$$m = -2$$

---

[23]When graphing by hand, it is best to select two points that are far apart on the graph. As an example, if the scale on the x-axis runs from -10 to 10, it is prudent to choose an x value in the vicinity of -10 and another in the vicinity of 10 and then find the corresponding y values of P_1 and P_2. The farther apart the points, the more accurate the graph will be. However, when using technological tools to graph a straight line, it does not matter how far apart the points are.

Graph: To graph the line, we choose the x- and y-intercept, i.e., $(0,0)$, as one point and the point $(1,-2)$ found earlier as our second point.

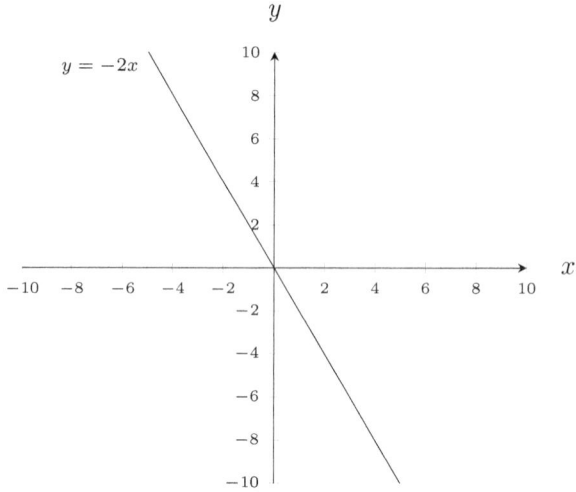

c. Formal form:

$$y \; = \; 1x$$

x-intercept: $(0,0)$
y-intercept: $(0,0)$

Slope: We choose the x- and y-intercept, i.e., $(0,0)$, as P_1. For P_2, we set $x = 1$ and find the corresponding y value using the equation $y = x$:

$$y \; = \; x$$
$$y \; = \; 1$$

This shows that the point $(1,1)$ is on the line. We choose this as P_2. The slope of the line is

$$m \; = \; \frac{y_2 \; - \; y_1}{x_2 \; - \; x_1}$$
$$m \; = \; \frac{1 \; - \; 0}{1 \; - \; 0}$$
$$m \; = \; \frac{1}{1}$$
$$m \; = \; 1$$

Graph: To graph the line, we choose the x- and y-intercept, i.e., $(0,0)$, as one point and the point $(1,1)$ found earlier as our second point.

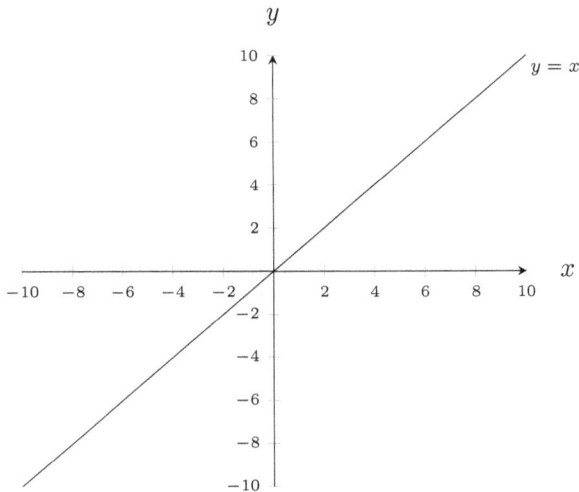

d. Formal form:

$$x = 3y$$
$$3y = x$$
$$y = \frac{1}{3}x$$

x-intercept: $(0,0)$
y-intercept: $(0,0)$

Slope: We choose the x- and y-intercept, i.e., $(0,0)$, as P_1. For P_2, we set $x = 3$ and find the corresponding y value using the equation $y = \frac{1}{3}x$:

$$y = \frac{1}{3}x$$
$$y = \frac{1}{3} \times 3$$
$$y = 1$$

This shows that the point $(3,1)$ is on the line. We choose this as P_2. The slope of the line is

$$m = \frac{y_2 - y_1}{x_2 - x_1}$$
$$m = \frac{1 - 0}{3 - 0}$$
$$m = \frac{1}{3}$$

Graph: To graph the line, we choose the x- and y-intercept, i.e., $(0,0)$, as one point and the point $(3,1)$ found earlier as our second point.

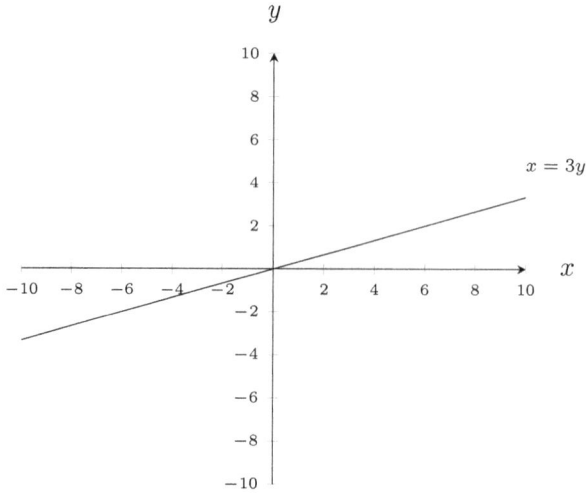

e. Formal form:

$$y = \frac{2}{5}x$$

x-intercept: $(0,0)$
y-intercept: $(0,0)$

Slope: We choose the x- and y-intercept, i.e., $(0,0)$, as P_1. For P_2, we set $x = 5$ and find the corresponding y value using the equation $y = \frac{2}{5}x$:

$$y = \frac{2}{5}x$$

$$y = \frac{2}{5} \times 5$$

$$y = 2$$

This shows that the point $(5,2)$ is on the line. We choose this as P_2. The slope of the line is

$$m = \frac{y_2 - y_1}{x_2 - x_1}$$

$$m = \frac{2 - 0}{5 - 0}$$

$$m = \frac{2}{5}$$

Graph: To graph the line, we choose the x- and y-intercept, i.e., $(0,0)$, as one point and the point $(5,2)$ found earlier as our second point.

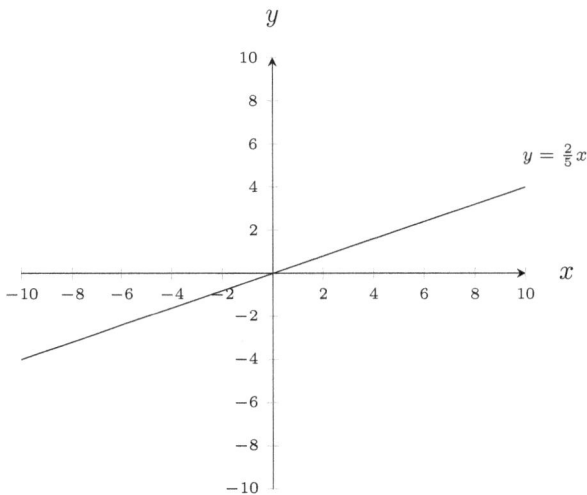

f. Formal form:

$$\frac{y}{x} = 4$$

$$y = 4x$$

x-intercept: $(0,0)$
y-intercept: $(0,0)$

Slope: We choose the x- and y-intercept, i.e., $(0,0)$, as P_1. For P_2, we set $x = 1$ and find the corresponding y value using the equation $y = 4x$:

$$y = 4x$$

$$y = 4 \times 1$$

$$y = 4$$

This shows that the point $(1,4)$ is on the line. We choose this as P_2. The slope of the line is

$$m = \frac{y_2 - y_1}{x_2 - x_1}$$

$$m = \frac{4 - 0}{1 - 0}$$

$$m = \frac{4}{1}$$

$$m = 4$$

Graph: To graph the line, we choose the x- and y-intercept, i.e., $(0,0)$, as one point and the point $(1,4)$ found earlier as our second point.

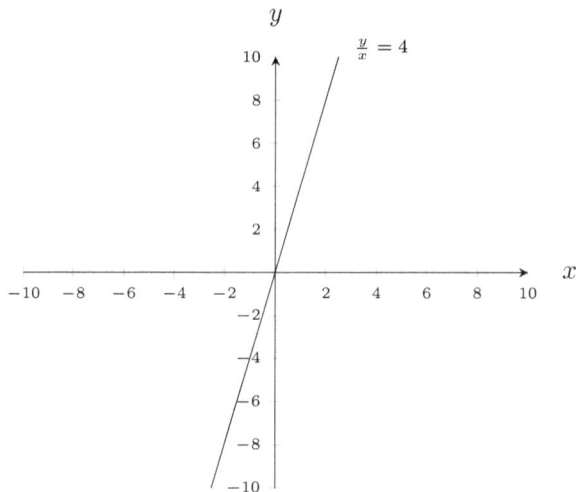

g. Formal form:

$$\frac{x}{y} = -1$$

$$y = \frac{x}{-1}$$

$$y = -x$$

$$y = -1x$$

x-intercept: $(0,0)$
y-intercept: $(0,0)$

Slope: We choose the x- and y-intercept, i.e., $(0,0)$, as P_1. For P_2, we set $x = 1$ and find the corresponding y value using the equation $y = -x$:

$$y = -x$$

$$y = -1$$

This shows that the point $(1,-1)$ is on the line. We choose this as P_2. The slope of the line is

$$m = \frac{y_2 - y_1}{x_2 - x_1}$$

$$m = \frac{-1 - 0}{1 - 0}$$

$$m = \frac{-1}{1}$$

$$m = -1$$

Graph: To graph the line, we choose the x- and y-intercept, i.e., $(0,0)$, as one point and the point $(1,-1)$ found earlier as our second point.

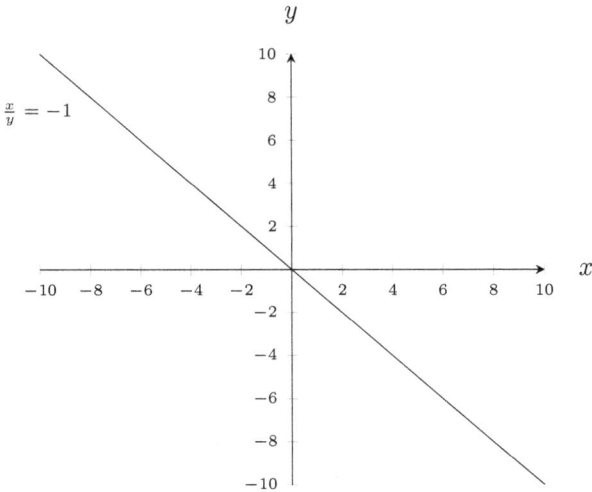

h. Formal form:

$$4y - x = 0$$
$$4y = 0 + x$$
$$4y = x$$
$$y = \frac{1}{4}x$$

x-intercept: $(0,0)$
y-intercept: $(0,0)$

Slope: We choose the x- and y-intercept, i.e., $(0,0)$, as P_1. For P_2, we set $x = 4$ and find the corresponding y value using the equation $y = \frac{1}{4}x$:

$$y = \frac{1}{4}x$$
$$y = \frac{1}{4} \times 4$$
$$y = 1$$

This shows that the point $(4,1)$ is on the line. We choose this as P_2. The slope of the line is

$$m = \frac{y_2 - y_1}{x_2 - x_1}$$
$$m = \frac{1 - 0}{4 - 0}$$
$$m = \frac{1}{4}$$

Graph: To graph the line, we choose the x- and y-intercept, i.e., $(0,0)$, as one point and the point $(4,1)$ found earlier as our second point.

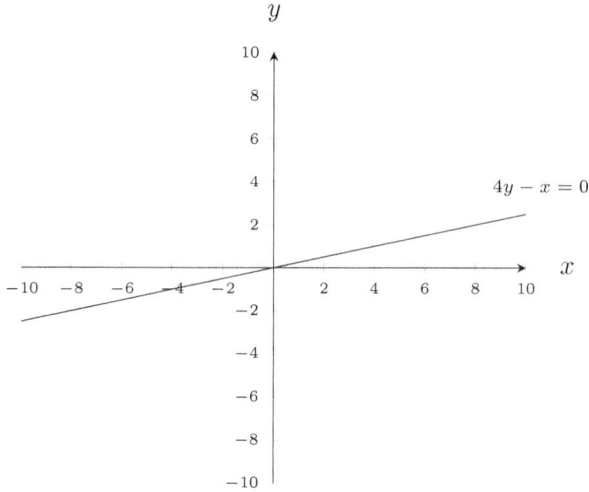

i. Formal form:

$$y - x = 0$$
$$y = 0 + x$$
$$y = x$$
$$y = 1x$$

x-intercept: $(0,0)$
y-intercept: $(0,0)$

Slope: We choose the x- and y-intercept, i.e., $(0,0)$, as P_1. For P_2, we set $x = 1$ and find the corresponding y value using the equation $y = x$:

$$y = x$$
$$y = 1$$

This shows that the point $(1,1)$ is on the line. We choose this as P_2. The slope of the line is

$$m = \frac{y_2 - y_1}{x_2 - x_1}$$
$$m = \frac{1 - 0}{1 - 0}$$
$$m = \frac{1}{1}$$
$$m = 1$$

Graph: To graph the line, we choose the x- and y-intercept, i.e., $(0,0)$, as one point and the point $(1,1)$ found earlier as our second point.

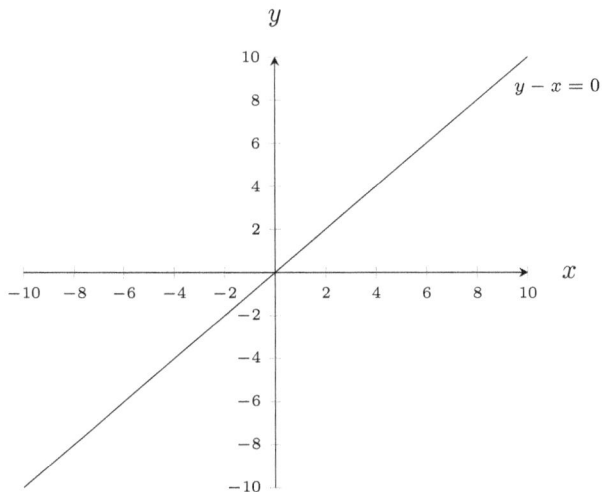

j. Formal form:

$$x - y = 0$$
$$-y = 0 - x$$
$$-y = -x$$
$$y = x$$
$$y = 1x$$

x-intercept: $(0,0)$
y-intercept: $(0,0)$

Slope: We choose the x- and y-intercept, i.e., $(0,0)$, as P_1. For P_2, we set $x = 1$ and find the corresponding y value using the equation $y = x$:

$$y = x$$
$$y = 1$$

This shows that the point $(1,1)$ is on the line. We choose this as P_2. The slope of the line is

$$m = \frac{y_2 - y_1}{x_2 - x_1}$$
$$m = \frac{1 - 0}{1 - 0}$$
$$m = \frac{1}{1}$$
$$m = 1$$

Graph: To graph the line, we choose the x- and y-intercept, i.e., $(0,0)$, as one point and the point $(1,1)$ found earlier as our second point.

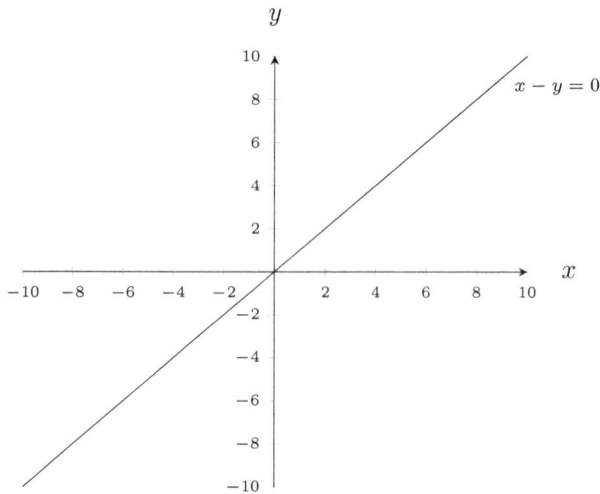

k. Formal form:

$$-\frac{y}{x} = \sqrt{2}$$
$$y = -\sqrt{2}x$$

x-intercept: $(0,0)$
y-intercept: $(0,0)$

Slope: We choose the x- and y-intercept, i.e., $(0,0)$, as P_1. For P_2, we set $x = 1$ and find the corresponding y value using the equation $y = -\sqrt{2}x$:

$$y = -\sqrt{2}x$$
$$y = -\sqrt{2} \times 1$$
$$y = -\sqrt{2}$$

This shows that the point $(1, -\sqrt{2})$ is on the line. We choose this as P_2. The slope of the line is

$$m = \frac{y_2 - y_1}{x_2 - x_1}$$
$$m = \frac{-\sqrt{2} - 0}{1 - 0}$$
$$m = \frac{-\sqrt{2}}{1}$$
$$m = -\sqrt{2}$$

Graph: To graph the line, we choose the x- and y-intercept, i.e., $(0,0)$, as one point and the point $\left(1, -\sqrt{2}\right)$ found earlier as our second point.[24]

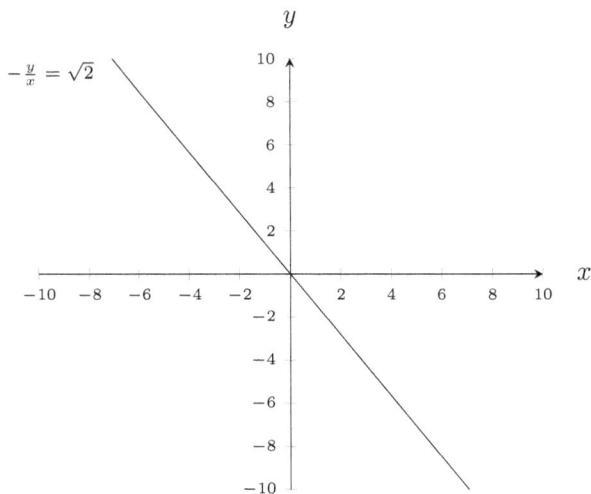

1. Formal form:

$$x = -y$$
$$y = -x$$
$$y = -1x$$

x-intercept: $(0,0)$
y-intercept: $(0,0)$

Slope: We choose the x- and y-intercept, i.e., $(0,0)$, as P_1. For P_2, we set $x = 1$ and find the corresponding y value using the equation $y = -x$:

$$y = -x$$
$$y = -1$$

This shows that the point $(1, -1)$ is on the line. We choose this as P_2. The slope of the line is

$$m = \frac{y_2 - y_1}{x_2 - x_1}$$
$$m = \frac{-1 - 0}{1 - 0}$$
$$m = \frac{-1}{1}$$
$$m = -1$$

---

[24]You can approximate the location of $\sqrt{2}$ on the number line by converting it to a decimal.

Graph: To graph the line, we choose the x- and y-intercept, i.e., $(0,0)$, as one point and the point $(1,-1)$ found earlier as our second point.

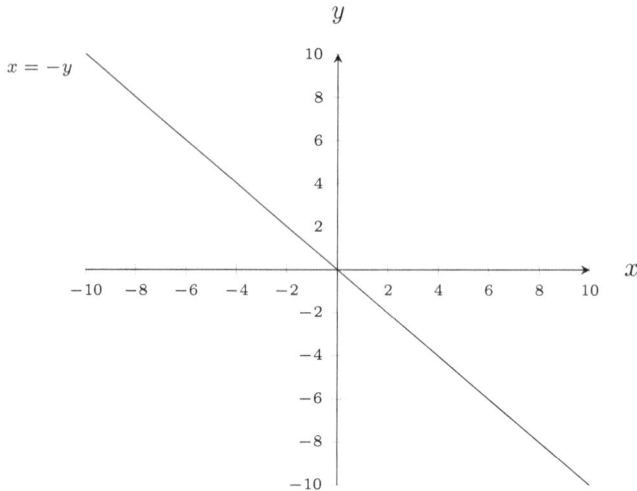

Exercise Set 12.2.1.4

1. It means that scaling the value of one of the quantities forces a scaling of the value of the other quantity with the value of one quantity increasing in size and the value of the other quantity decreasing in size.

2. It means that scaling the value of one of the quantities by a certain factor forces a corresponding scaling of the value of the other quantity by the inverse of the same factor.

3. $y = \frac{k}{x}$, $k \in \mathbb{R}, k \neq 0$

4. $yx = k$, $k \in \mathbb{R}, k \neq 0$

5. It means that scaling the value of either quantity by a certain factor forces a scaling of the value of the other quantity by the inverse of the same factor.

6. There is no simple relationship between change in the values of the quantities that are related through inverse proportion. However, we can state that $\Delta y = k\Delta\frac{1}{x}$ where x and y represent the values of the two quantities and k is the constant of proportionality.

7. The shape of the graph of the model for a relationship of type inverse proportion is a double curve with a discontinuity at the y-axis and another at the x-axis.

8. The effect of k in $y = \frac{k}{x}$, $k \in \mathbb{R}$, $k < 0$ is to move the two branches of the graph toward and away from the origin. The higher the size of k, the farther away the graph of the equation is from the origin. The following figure shows the graphs of three equations with negative values of k: $-\frac{1}{2}$, -1 and -2, with

corresponding equations $y = -\frac{1}{2x}$, $y = -\frac{1}{x}$ and $y = -\frac{2}{x}$. Note that as the size of k increases, the graph moves away from the origin. Furthermore, when k is 0 the graph maps onto the two axes.

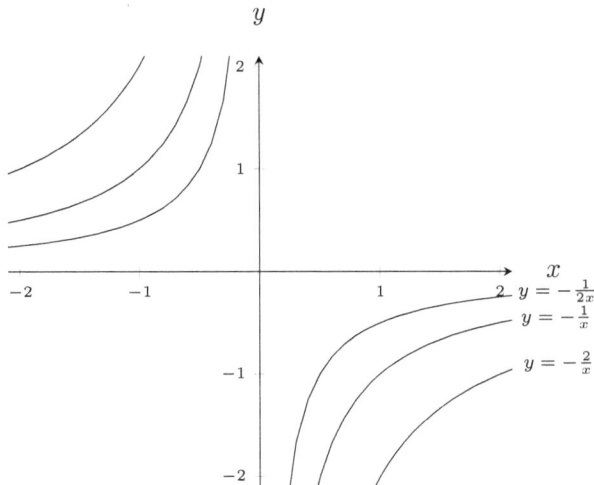

9. a. $y = \frac{2}{x}$ is a model for a relationship of type inverse proportion as it has the form of $y = \frac{k}{x}$, $k \in \mathbb{R}$, $k \neq 0$ with $k = 2$.

   b. $y = \frac{-6}{x}$ is a model for a relationship of type inverse proportion as it has the form of $y = \frac{k}{x}$, $k \in \mathbb{R}$, $k \neq 0$ with $k = -6$.

   c. $y = \frac{1}{x}$ is a model for a relationship of type inverse proportion as it has the form of $y = \frac{k}{x}$, $k \in \mathbb{R}$, $k \neq 0$ with $k = 1$.

   d. $y = \frac{2}{5x}$ may be rewritten as $\frac{\frac{2}{5}}{x}$ which shows that it is a model for a relationship of type inverse proportion as it has the form of $y = \frac{k}{x}$, $k \in \mathbb{R}$, $k \neq 0$ with $k = \frac{2}{5}$.

   e. $y = -\frac{1}{3x}$ may be rewritten as $\frac{-\frac{1}{3}}{x}$ which shows that it is a model for a relationship of type inverse proportion as it has the form of $y = \frac{k}{x}$, $k \in \mathbb{R}$, $k \neq 0$ with $k = -\frac{1}{3}$.

   f. Rearranging the equation $yx = 7$ to $y = \frac{7}{x}$ shows that it is a model for a relationship of type inverse proportion as it has the form of $y = \frac{k}{x}$, $k \in \mathbb{R}$, $k \neq 0$ with $k = 7$.[25]

---

[25] We can also compare equations to the conservation form of the model for inverse proportion. As an example, we can argue that the equation $yx = 7$ is a model for relationships of type inverse proportion since it matches the conservation form of the equation for inverse proportion, i.e., $yx = k$, $k \in \mathbb{R}$, $k \neq 0$.

While it is always possible to compare a given equation to the conservation form, for the sake of uniformity we will use the formal form for such comparisons but we do advise the reader to always look for this alternative line of reasoning.

g. $yx = -\frac{3}{4}$ may be rewritten as $y = \frac{-\frac{3}{4}}{x}$ and is therefore a model for a relationship of type inverse proportion as it has the form of $y = \frac{k}{x}$, $k \in \mathbb{R}$, $k \neq 0$ with $k = -\frac{3}{4}$.

h. $yx = -3$ may be rewritten as $y = \frac{-3}{x}$ and is therefore a model for a relationship of type inverse proportion as it has the form of $y = kx$, $k \in \mathbb{R}$, $k \neq 0$ with $k = -3$.

i. $\frac{y}{x} = \frac{1}{5}$ is not a model for a relationship of type inverse proportion as it does not have the form of $y = kx$, $k \in \mathbb{R}$, $k \neq 0$ even if it is written as $y = \frac{1}{5}x$. Note that in the formal form x is being divided by on the right side of the equation whereas in $y = \frac{1}{5}x$, x is being multiplied by on the right side of the equation.

j. $\frac{x}{y} = -1$ may be rewritten as $y = -x$:

$$\frac{x}{y} = -1$$

$$y = \frac{x}{-1}$$

$$y = -x$$

This may be seen as $y = -1x$ and is therefore not a model for a relationship of type inverse proportion as it does not have the form of $y = \frac{k}{x}$, $k \in \mathbb{R}$, $k \neq 0$. Note that in the formal form x is being divided by on the right side of the equation whereas in $y = -1x$, x is being multiplied by on the right side of the equation.

k. $\frac{x}{y} = 4$ may be rewritten as $y = \frac{1}{4}x$:

$$\frac{x}{y} = 4$$

$$y = \frac{x}{4}$$

$$y = \frac{1}{4}x$$

and is therefore not a model for a relationship of type inverse proportion as it does not have the form of $y = \frac{k}{x}$, $k \in \mathbb{R}$, $k \neq 0$. Note that in the formal form x is being divided by on the right side of the equation whereas in $y = \frac{1}{4}x$, x is being multiplied by on the right side of the equation.

l. $y = \frac{8}{x}$ is a model for a relationship of type inverse proportion as it has the form of $y = \frac{k}{x}$, $k \in \mathbb{R}$, $k \neq 0$ with $k = 8$.

m. $y = -\frac{2}{x}$ may be rewritten as $y = \frac{-2}{x}$ and is therefore a model for a relationship of type inverse proportion as it has the form of $y = \frac{k}{x}$, $k \in \mathbb{R}$, $k \neq 0$ with $k = -2$.

n. $y = 3x$ is not a model for a relationship of type inverse proportion as it does not have the form of $y = \frac{k}{x}$, $k \in \mathbb{R}$, $k \neq 0$. Note that in the formal form x is being divided by on the right side of the equation whereas in $y = 3x$, x is being multiplied by on the right side of the equation.

o. $y = -x$ may be seen as $y = -1x$ and is, therefroe, not a model for a relationship of type inverse proportion as it does not have the form of $y = \frac{k}{x}$, $k \in \mathbb{R}$, $k \neq 0$. Note that in the formal form x is being divided by on the right side of the equation whereas in $y = -x$ or its equivalent statement $y = -1x$, x is being multiplied by on the right side of the equation.

p. $x = \frac{-2}{y}$ may be written as $y = \frac{-2}{x}$ and is, therefroe, a model for a relationship of type inverse proportion as it has the form of $y = \frac{k}{x}$, $k \in \mathbb{R}$, $k \neq 0$.

q. $x = \frac{1}{6}y$ may be written as $y = 6x$ and is, therefroe, not a model for a relationship of type inverse proportion as it does not have the form of $y = \frac{k}{x}$, $k \in \mathbb{R}$, $k \neq 0$. Note that in the formal form x is being divided by on the right side of the equation whereas in $y = 6x$, x is being multiplied by on the right side of the equation.

r. $-2x = 5y$ may be written as $y = -\frac{2}{5}x$ and is, therefroe, not a model for a relationship of type inverse proportion as it does not have the form of $y = \frac{k}{x}$, $k \in \mathbb{R}$, $k \neq 0$. Note that in the formal form x is being divided by on the right side of the equation whereas in $y = -\frac{2}{5}x$, x is being multiplied by on the right side of the equation.

s. $3y = 6x$ may be written as $y = 2x$:

$$3y = 6x$$
$$y = \frac{6}{3}x$$
$$y = 2x$$

and is, therefroe, not a model for a relationship of type inverse proportion as it does not have the form of $y = \frac{k}{x}$, $k \in \mathbb{R}$, $k \neq 0$. Note that in the formal form x is being divided by on the right side of the equation whereas in $y = 2x$, x is being multiplied by on the right side of the equation.

t. $y = -8 + \frac{5}{x}$ is not a model for a relationship of type inverse proportion as it does not have the form of $y = \frac{k}{x}$, $k \in \mathbb{R}$, $k \neq 0$. Note that in the formal form there is only one term on the right side of the equation whereas in $y = -8 + \frac{5}{x}$ there are two terms on the right side of the equation.

u. $3x - 1 = y$ or its equivalent $y = 3x - 1$ is not a model for a relationship of type inverse proportion as it does not have the form of $y = \frac{k}{x}$, $k \in \mathbb{R}$, $k \neq 0$. Note that in the formal form there is only one term on the right side of the equation whereas in $y = 3x - 1$ there are two terms on the right side of the equation.

v. $-2x - \frac{3}{y} = 0$ may be rewritten as $y = \frac{-\frac{3}{2}}{x}$:

$$-2x - \frac{3}{y} = 0$$

$$-\frac{3}{y} = 0 + 2x$$

$$-\frac{3}{y} = 2x$$

$$y = -\frac{3}{2x}$$

$$y = \frac{-\frac{3}{2}}{x}$$

and is therefore a model for a relationship of type inverse proportion as it has the form of $y = \frac{k}{x}$, $k \in \mathbb{R}$, $k \neq 0$ with $k = -\frac{3}{2}$.

w. $7x + 2y = 4$ may be rewritten as $y = 2 - \frac{7}{2}x$:

$$7x + 2y = 4$$

$$2y = 4 - 7x$$

$$y = \frac{1}{2}(4 - 7x)$$

$$y = 2 - \frac{7}{2}x$$

and is, therefore, not a model for a relationship of type inverse proportion as it does not have the form of $y = \frac{k}{x}$, $k \in \mathbb{R}$, $k \neq 0$. Note that in the formal form there is only one term on the right side of the equation whereas in $y = 2 - \frac{7}{2}x$ there are two terms on the right side of the equation.

10. a. Formal form:

$$y = \frac{7}{x}$$

Conservation form:

$$y = \frac{7}{x}$$

$$yx = 7$$

b. Formal form:

$$y = \frac{-2}{x}$$

Conservation form:

$$y = \frac{-2}{x}$$

$$yx = -2$$

c. Formal form:

$$y = \frac{1}{x}$$

Conservation form:

$$y = \frac{1}{x}$$

$$yx = 1$$

d. Formal form:

$$x = \frac{3}{y}$$

$$y = \frac{3}{x}$$

Conservation form:

$$x = \frac{3}{y}$$

$$yx = 3$$

e. Formal form:

$$y = \frac{2}{5x}$$

$$y = \frac{\frac{2}{5}}{x}$$

Conservation form:

$$y = \frac{2}{5x}$$

$$yx = \frac{2}{5}$$

f. Formal form:

$$yx = 4$$

$$y = \frac{4}{x}$$

Conservation form:

$$yx = 4$$

g. Formal form:

$$yx = -1$$

$$y = \frac{-1}{x}$$

Conservation form:

$$yx = -1$$

h. Formal form:

$$4y - \frac{1}{x} = 0$$

$$4y = 0 + \frac{1}{x}$$

$$4y = \frac{1}{x}$$

$$y = \frac{1}{4x}$$

$$y = \frac{\frac{1}{4}}{x}$$

Conservation form:

$$4y - \frac{1}{x} = 0$$

$$4y = 0 + \frac{1}{x}$$

$$4y = \frac{1}{x}$$

$$yx = \frac{1}{4}$$

i. Formal form:

$$\frac{1}{y} - x = 0$$

$$\frac{1}{y} = 0 + x$$

$$\frac{1}{y} = x$$

$$y = \frac{1}{x}$$

Conservation form:

$$\frac{1}{y} - x = 0$$

$$\frac{1}{y} = 0 + x$$

$$\frac{1}{y} = x$$

$$yx = 1$$

j. Formal form:

$$x - \frac{1}{y} = 0$$

$$x = 0 + \frac{1}{y}$$

$$x = \frac{1}{y}$$

$$y = \frac{1}{x}$$

Conservation form:

$$x - \frac{1}{y} = 0$$

$$x = 0 + \frac{1}{y}$$

$$x = \frac{1}{y}$$

$$yx = 1$$

k. Formal form:

$$xy = \sqrt{2}$$

$$y = \frac{\sqrt{2}}{x}$$

Conservation form:

$$xy = \sqrt{2}$$

$$yx = \sqrt{2}$$

1. Formal form:

$$x = -\frac{1}{y}$$

$$y = -\frac{1}{x}$$

$$y = \frac{-1}{x}$$

Conservation form:

$$x = -\frac{1}{y}$$

$$yx = -1$$

11. In each solution below, we use our earlier finding that the graph of an equation of type inverse proportion has no x- or y-intercepts. We use the values -4, -2, -1, $-\frac{1}{2}$, $-\frac{1}{4}$ as well as $\frac{1}{4}$, $\frac{1}{2}$, 1, 2 and 4 to graph the two branches of the graph. For each value, the associated y value is calculated using the relevant equation.[26]

a. Formal form:

$$y = \frac{7}{x}$$

There are no x- or y-intercepts.

Graph: To graph the line, we use the points $\left(-4, -\frac{7}{4}\right)$, $\left(-2, -\frac{7}{2}\right)$, $\left(-1, -7\right)$, $\left(-\frac{1}{2}, -14\right)$, $\left(-\frac{1}{4}, -28\right)$, as well as $\left(\frac{1}{4}, 28\right)$, $\left(\frac{1}{2}, 14\right)$, $\left(1, 7\right)$, $\left(2, \frac{7}{2}\right)$, $\left(4, \frac{7}{4}\right)$.

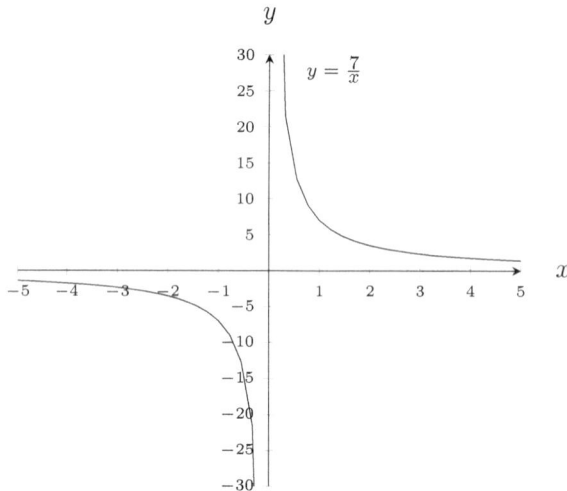

b. Formal form:

$$y = \frac{-2}{x}$$

There are no x- or y-intercepts.

---

[26] Here we only quote the y values and leave detailed calculation of these values to the reader.

Graph: To graph the line, we use the points $\left(-4, \frac{1}{2}\right)$, $(-2, 1)$, $(-1, 2)$, $\left(-\frac{1}{2}, 4\right)$, $\left(-\frac{1}{4}, 8\right)$, as well as $\left(\frac{1}{4}, -8\right)$, $\left(\frac{1}{2}, -4\right)$, $(1, -2)$, $(2, -1)$, $\left(4, -\frac{1}{2}\right)$.

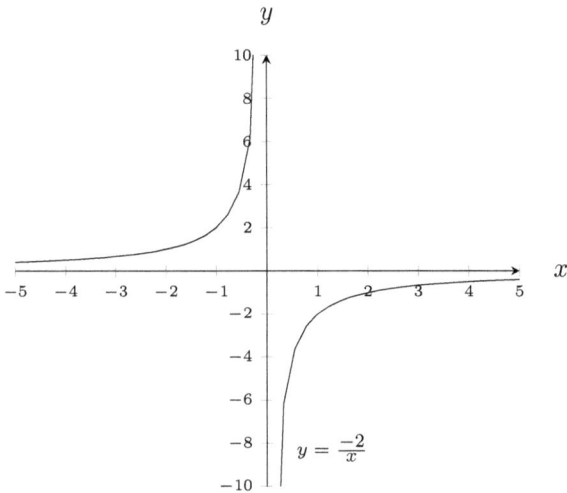

$$y = \frac{-2}{x}$$

c. Formal form:

$$y = \frac{1}{x}$$

There are no x- or y-intercepts.

Graph: To graph the line, we use the points $\left(-4, -\frac{1}{4}\right)$, $\left(-2, -\frac{1}{2}\right)$, $(-1, -1)$, $\left(-\frac{1}{2}, -2\right)$, $\left(-\frac{1}{4}, -4\right)$, as well as $\left(\frac{1}{4}, 4\right)$, $\left(\frac{1}{2}, 2\right)$, $(1, 1)$, $\left(2, \frac{1}{2}\right)$, $\left(4, \frac{1}{4}\right)$.

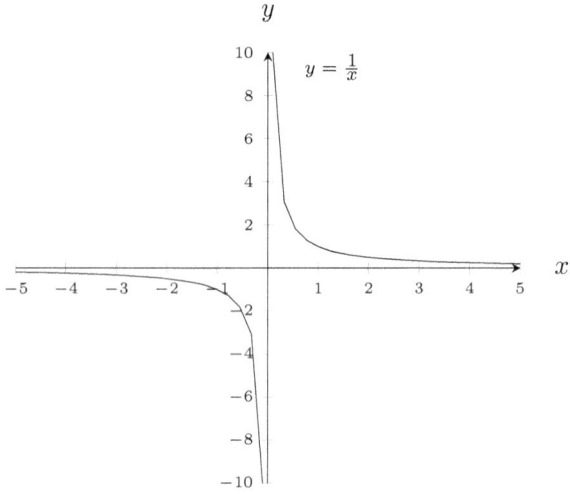

d. Formal form:

$$x = \frac{3}{y}$$

$$y = \frac{3}{x}$$

There are no x- or y-intercepts.

Graph: To graph the line, we use the points $\left(-4, -\frac{3}{4}\right)$, $\left(-2, -\frac{3}{2}\right)$, $\left(-1, -3\right)$, $\left(-\frac{1}{2}, -6\right)$, $\left(-\frac{1}{4}, -12\right)$, as well as $\left(\frac{1}{4}, 12\right)$, $\left(\frac{1}{2}, 6\right)$, $(1, 3)$, $\left(2, \frac{3}{2}\right)$, $\left(4, \frac{3}{4}\right)$.

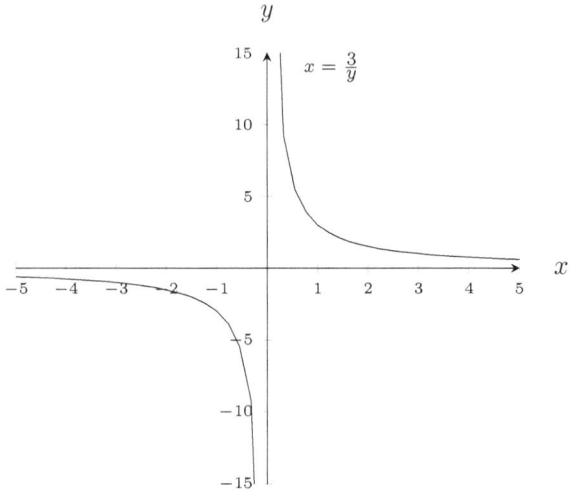

e. Formal form:

$$y = \frac{2}{5x}$$

$$y = \frac{\frac{2}{5}}{x}$$

There are no x- or y-intercepts.

Graph: To graph the line, we use the points $\left(-4, -\frac{1}{10}\right)$, $\left(-2, -\frac{1}{5}\right)$, $\left(-1, -\frac{2}{5}\right)$, $\left(-\frac{1}{2}, -\frac{4}{5}\right)$, $\left(-\frac{1}{4}, -\frac{8}{5}\right)$, as well as $\left(\frac{1}{4}, \frac{8}{5}\right)$, $\left(\frac{1}{2}, \frac{4}{5}\right)$, $\left(1, \frac{2}{5}\right)$, $\left(2, \frac{1}{5}\right)$, $\left(4, \frac{1}{10}\right)$.

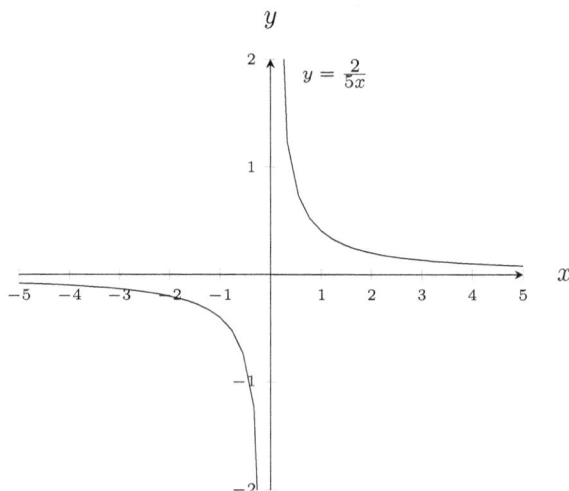

f. Formal form:

$$yx = 4$$

$$y = \frac{4}{x}$$

There are no x- or y-intercepts.

Graph: To graph the line, we use the points $(-4, -1)$, $(-2, -2)$, $(-1, -4)$, $\left(-\frac{1}{2}, -8\right)$, $\left(-\frac{1}{4}, -16\right)$, as well as $\left(\frac{1}{4}, 16\right)$, $\left(\frac{1}{2}, 8\right)$, $(1, 4)$, $(2, 2)$, $(4, 1)$.

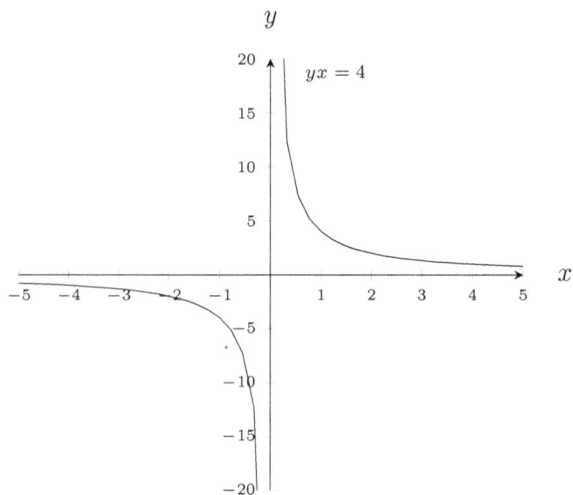

g. Formal form:

$$yx = -1$$
$$y = \frac{-1}{x}$$

There are no x- or y-intercepts.

Graph: To graph the line, we use the points $\left(-4, \frac{1}{4}\right)$, $\left(-2, \frac{1}{2}\right)$, $(-1, 1)$, $\left(-\frac{1}{2}, 2\right)$, $\left(-\frac{1}{4}, 4\right)$, as well as $\left(\frac{1}{4}, -4\right)$, $\left(\frac{1}{2}, -2\right)$, $(1, -1)$, $\left(2, -\frac{1}{2}\right)$, $\left(4, -\frac{1}{4}\right)$.

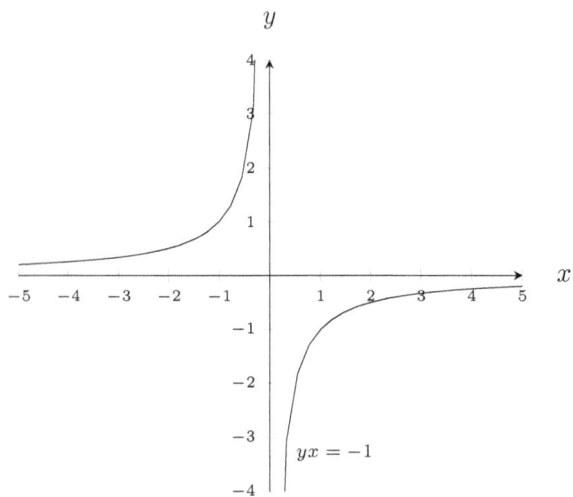

h. Formal form:

$$4y - \frac{1}{x} = 0$$

$$4y = 0 + \frac{1}{x}$$

$$4y = \frac{1}{x}$$

$$y = \frac{1}{4x}$$

$$y = \frac{\frac{1}{4}}{x}$$

There are no x- or y-intercepts.

Graph: To graph the line, we use the points $\left(-4, -\frac{1}{16}\right)$, $\left(-2, -\frac{1}{8}\right)$, $\left(-1, -\frac{1}{4}\right)$, $\left(-\frac{1}{2}, -\frac{1}{2}\right)$, $\left(-\frac{1}{4}, -1\right)$, as well as $\left(\frac{1}{4}, 1\right)$, $\left(\frac{1}{2}, \frac{1}{2}\right)$, $\left(1, \frac{1}{4}\right)$, $\left(2, \frac{1}{8}\right)$, $\left(4, \frac{1}{16}\right)$.

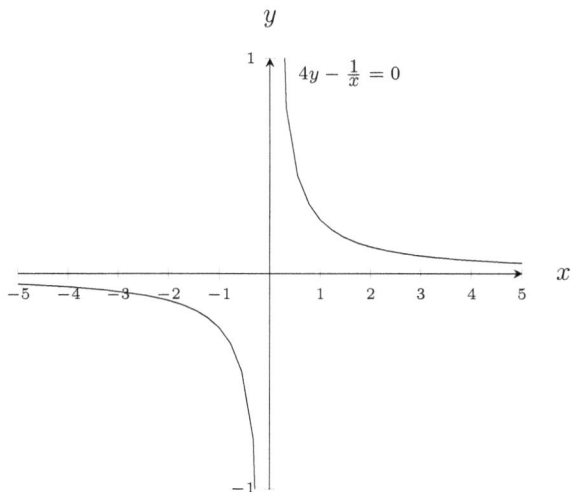

i. Formal form:

$$\frac{1}{y} - x = 0$$

$$\frac{1}{y} = 0 + x$$

$$\frac{1}{y} = x$$

$$y = \frac{1}{x}$$

There are no x- or y-intercepts.

Graph: To graph the line, we use the points $\left(-4, -\frac{1}{4}\right)$, $\left(-2, -\frac{1}{2}\right)$, $(-1, -1)$, $\left(-\frac{1}{2}, -2\right)$, $\left(-\frac{1}{4}, -4\right)$, as well as $\left(\frac{1}{4}, 4\right)$, $\left(\frac{1}{2}, 2\right)$, $(1, 1)$, $\left(2, \frac{1}{2}\right)$, $\left(4, \frac{1}{4}\right)$.

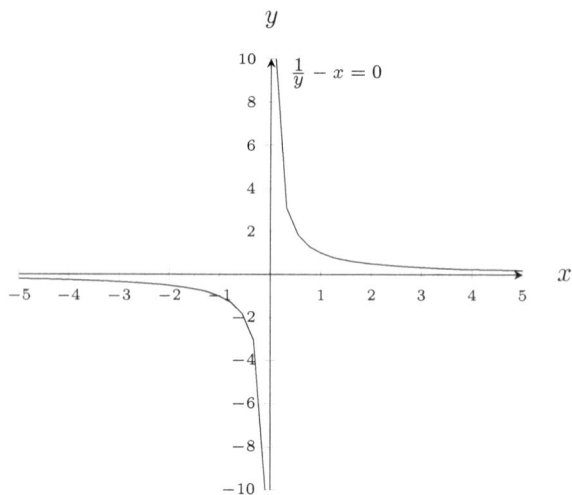

j. Formal form:

$$x - \frac{1}{y} = 0$$

$$-\frac{1}{y} = 0 - x$$

$$-\frac{1}{y} = -x$$

$$\frac{1}{y} = x$$

$$y = \frac{1}{x}$$

There are no x- or y-intercepts.

Graph: To graph the line, we use the points $\left(-4, -\frac{1}{4}\right)$, $\left(-2, -\frac{1}{2}\right)$, $(-1, -1)$, $\left(-\frac{1}{2}, -2\right)$, $\left(-\frac{1}{4}, -4\right)$, as well as $\left(\frac{1}{4}, 4\right)$, $\left(\frac{1}{2}, 2\right)$, $(1, 1)$, $\left(2, \frac{1}{2}\right)$, $\left(4, \frac{1}{4}\right)$.

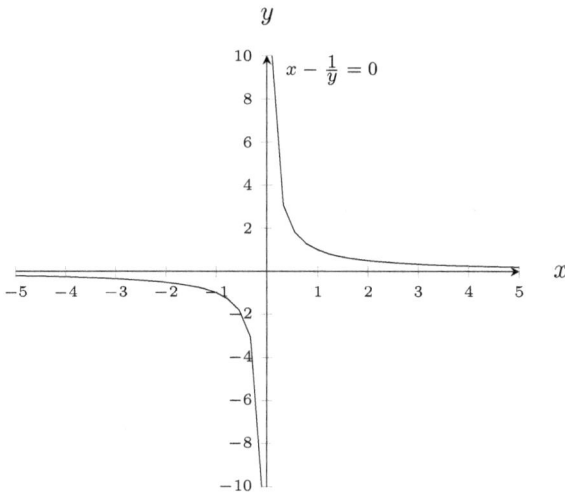

k. Formal form:

$$xy = \sqrt{2}$$

$$y = \frac{\sqrt{2}}{x}$$

There are no x- or y-intercepts.

Graph: To graph the line, we use the points $\left(-4, -\frac{1}{4}\sqrt{2}\right)$, $\left(-2, -\frac{1}{2}\sqrt{2}\right)$, $\left(-1, -\sqrt{2}\right)$, $\left(-\frac{1}{2}, -2\sqrt{2}\right)$, $\left(-\frac{1}{4}, -4\sqrt{2}\right)$, as well as $\left(\frac{1}{4}, 4\sqrt{2}\right)$, $\left(\frac{1}{2}, 2\sqrt{2}\right)$, $\left(1, \sqrt{2}\right)$, $\left(2, \frac{1}{2}\sqrt{2}\right)$, $\left(4, \frac{1}{4}\sqrt{2}\right)$.

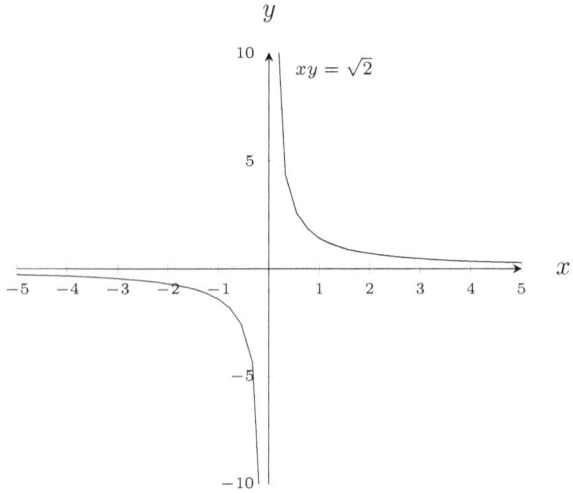

1. Formal form:

$$x = -\frac{1}{y}$$
$$y = \frac{-1}{x}$$

There are no x- or y-intercepts.

Graph: To graph the line, we use the points $\left(-4, \frac{1}{4}\right)$, $\left(-2, \frac{1}{2}\right)$, $\left(-1, 1\right)$, $\left(-\frac{1}{2}, 2\right)$, $\left(-\frac{1}{4}, 4\right)$, as well as $\left(\frac{1}{4}, -4\right)$, $\left(\frac{1}{2}, -2\right)$, $\left(1, -1\right)$, $\left(2, -\frac{1}{2}\right)$, $\left(4, -\frac{1}{4}\right)$.

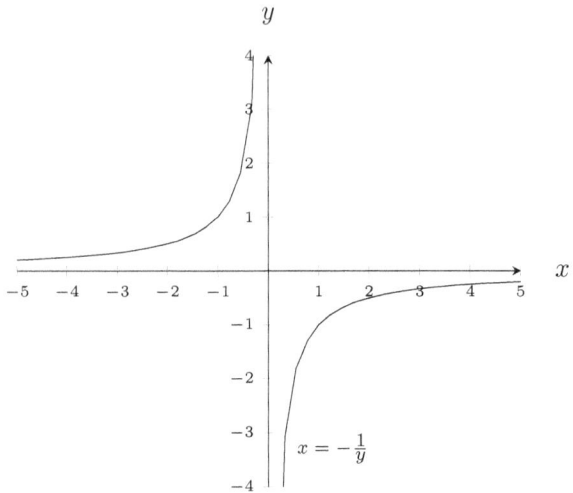

Exercise Set 12.2.1.5

1. a. Formal model: $y = x^k$
   Conservation model: $\log_x y = k$

   b. Formal model: $y = \sqrt[k]{x}$
   Conservation model: $\log_x y = \frac{1}{k}$

   c. Formal model: $y = k^x$
   Conservation model: $\frac{1}{x} \log_k y = 1$

   d. Formal model: $y = \log_k x$
   Conservation model: $\frac{1}{x} k^y = 1$

2. a. Direct exponential

   b. Direct power

   c. Inverse power

   d. Inverse exponential

3. a. There is no x-intercept. The y-intercept is $(0, 1)$. The graph is given below.

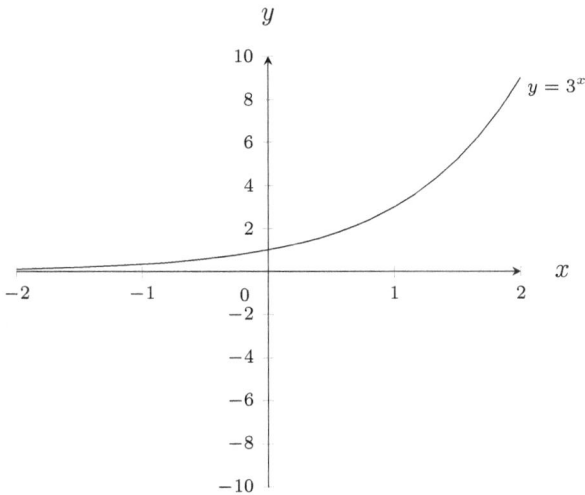

b. The x- and y-intercepts are both $(0,0)$. The graph is given below.

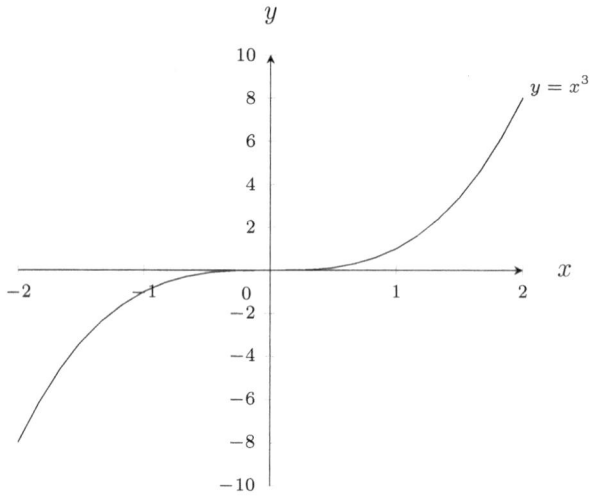

c. The x- and y-intercepts are both $(0,0)$. The graph is given below.

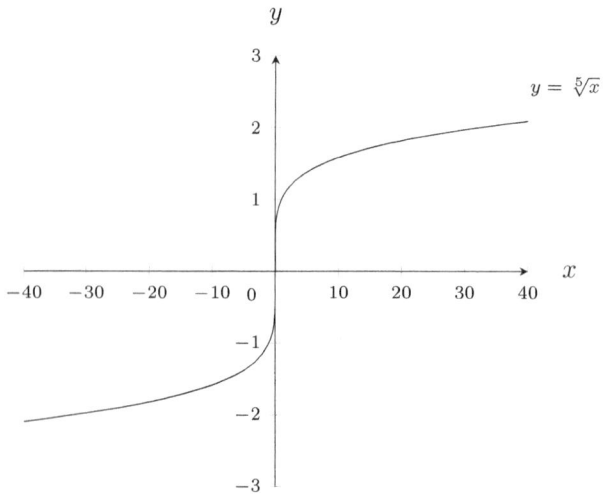

d. The x-intercept is $(1,0)$. There is no y-intercept. The graph is given below.

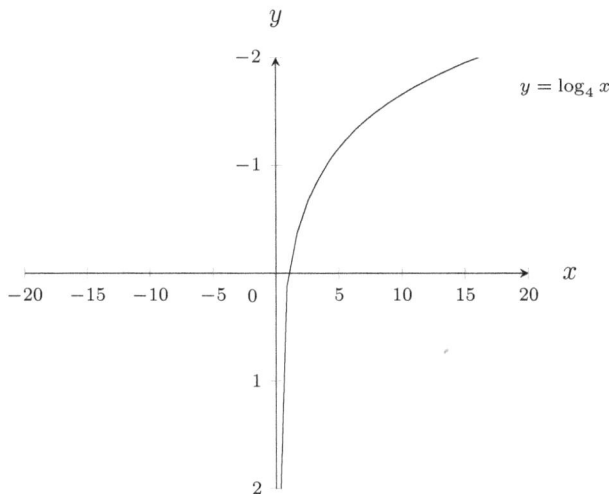

Exercise Set 12.2.2.1

1. $y = mx + b$, $m, b \in \mathbb{R}$, $m \neq 0$

2. It means that if the value of one quantity is linearly dependent on the value of a second quantity, then the value of the second quantity is linearly dependent on the value of the first quantity.

3. $\Delta y = m \Delta x$

4. The graph of the model for a linear relationship is a straight line

5. m

6. a. $y = 2x + 1$ is a model for a linear relationship as it has the form of $y = mx + b$ with $m = 2$ and $b = 1$.

   b. $y = -6x - 5$ may be viewed as $y = -6x + (-5)$ and is, therefore, a model for a linear relationship as it has the form of $y = mx + b$ with $m = -6$ and $b = -5$.

   c. $y = x + 3$ may be viewed as $y = 1x + 3$ and is, therefore, a model for a linear relationship as it has the form of $y = mx + b$ with $m = 1$ and $b = 3$.

   d. $y = \frac{2}{5}x - \frac{1}{4}$ may be viewed as $y = \frac{2}{5}x + \left(-\frac{1}{4}\right)$ and is, therefore, a model for a linear relationship as it has the form of $y = mx + b$ with $m = \frac{2}{5}$ and $b = -\frac{1}{4}$.

   e. $y = -x + 3$ may be viewed as $y = -1x + 3$ and is, therefore, a model for a linear relationship as it has the form of $y = mx + b$ with $m = -1$ and $b = 3$.

f. $y = x + 1$ may be viewed as $y = 1x + 1$ and is, therefore, a model for a linear relationship as it has the form of $y = mx + b$ with $m = 1$ and $b = 1$.

g. $y = 3x$ may be viewed as $y = 3x + 0$ and is, therefore, a model for a linear relationship as it has the form of $y = mx + b$ with $m = 3$ and $b = 0$.

h. $y = -3x$ may be viewed as $y = -3x + 0$ and is, therefore, a model for a linear relationship as it has the form of $y = mx + b$ with $m = -3$ and $b = 0$.

i. $y = 4$ may be viewed as $y = 0x + 4$ and is, therefore, a model for a linear relationship as it has the form of $y = mx + b$ with $m = 0$ and $b = 4$.

j. $y = x^2$ is not a model for a linear relationship as it does not have the form of $y = mx + b$. Note that there is no x^2 in the linear model $y = mx + b$.

k. $y = \sqrt{x} + 1$ is not a model for a linear relationship as it does not have the form of $y = mx + b$. Note that there is no $\sqrt{x}$ in the linear model $y = mx + b$.

l. $y = \frac{x}{10} + 3$ may be viewed as $y = \frac{1}{10}x + 3$ and is, therefore, a model for a linear relationship as it has the form of $y = mx + b$ with $m = \frac{1}{10}$ and $b = 3$.

m. $y = -\frac{x}{2}$ may be viewed as $y = -\frac{1}{2}x + 0$ and is, therefore, a model for a linear relationship as it has the form of $y = mx + b$ with $m = -\frac{1}{2}$ and $b = 0$.

n. $y = \frac{3}{x}$ is not a model for a linear relationship as it does not have the form of $y = mx + b$. Note that, in the linear model, $y = mx + b$, x is being multiplied by while in $y = \frac{3}{x}$, x is being divided by.

o. $y = -\frac{1}{x+1}$ is not a model for a linear relationship as it does not have the form of $y = mx + b$. Note that, in the linear model, $y = mx + b$, x is being multiplied by while in $y = -\frac{1}{x+1}$, x is part of a divisor.

p. $x = -4y$ may be viewed as $y = -\frac{1}{4}x + 0$:

$$x = -4y$$
$$-4y = x$$
$$y = -\frac{1}{4}x$$
$$y = -\frac{1}{4}x + 0$$

and is, therefore, a model for a linear relationship as it has the form of $y = mx + b$ with $m = -\frac{1}{4}$ and $b = 0$.

q. $x = \frac{1}{6}y$ may be viewed as $y = 6x + 0$:

$$x = \frac{1}{6}y$$
$$\frac{1}{6}y = x$$
$$y = 6x$$
$$y = 6x + 0$$

and is, therefore, a model for a linear relationship as it has the form of $y = mx + b$ with $m = 6$ and $b = 0$.

r. $-2x = 5y^2$ is not a model for a linear relationship even if it is written as $y = \pm\sqrt{-\frac{2}{5}x}$ as it does not have the form of $y = mx + b$. Note that in the linear model, i.e., $y = mx + b$, x is not part of an argument of a root.

s. $y = x^2 + 3x + 2$ is not a model for a linear relationship. Note that there is no x^2 in the linear model $y = mx + b$.

t. $2y = -8 + 5x$ may be viewed as $y = \frac{5}{2}x + (-4)$:

$$2y = -8 + 5x$$
$$2y = 5x - 8$$
$$y = \frac{1}{2}(5x - 8)$$
$$y = \frac{5}{2}x - 4$$
$$y = \frac{5}{2}x + (-4)$$

and is, therefore, a model for a linear relationship as it has the form of $y = mx + b$ with $m = \frac{5}{2}$ and $b = -4$.

u. $3x - 1 = y$ may be viewed as $y = 3x + (-1)$:

$$3x - 1 = y$$
$$y = 3x - 1$$
$$y = 3x + (-1)$$

and is, therefore, a model for a linear relationship as it has the form of $y = mx + b$ with $m = 3$ and $b = -1$.

v. $y = \log x$ is not a model for a linear relationship. Note that in the linear model, i.e., $y = mx + b$, x is not an argument of a logarithm.

w. $x^2 + 2xy - y^2$ is not a model for a linear relationship. Note that the linear model, i.e., $y = mx + b$, does not involve x^2, xy, or y^2.

7. a.

$$y = 7x + 3$$

b.

$$y = -2x - 1$$
$$y = -2x + (-1)$$

c.

$$y = x$$
$$y = 1x + 0$$

d.

$$x = 3y$$
$$3y = x$$
$$y = \frac{1}{3}x + 0$$

e.

$$y = \frac{3}{4}x$$
$$y = \frac{3}{4}x + 0$$

f.

$$\frac{y}{x} = 4$$

$$y = 4x$$

$$y = 4x + 0$$

g.

$$\frac{y}{2x} = -1$$

$$y = -2x$$

$$y = -2x + 0$$

h.

$$4y - x = 0$$

$$4y = x + 0$$

$$4y = x$$

$$y = \frac{1}{4}x$$

$$y = \frac{1}{4}x + 0$$

i.

$$x = 3y - 7$$

$$3y - 7 = x$$

$$3y = x + 7$$

$$y = \frac{1}{3}(x + 7)$$

$$y = \frac{1}{3}x + \frac{7}{3}$$

j.

$$x + y = 0$$

$$y = -x + 0$$

$$y = -1x + 0$$

k.

$$-\frac{y}{x} = -6$$

$$y = -6(-x)$$

$$y = 6x$$

$$y = 6x + 0$$

l.

$$x = -y$$

$$-y = x$$

$$y = -x$$

$$y = -1x + 0$$

m.

$$y = 2(x - 2)$$

$$y = 2x - 4$$

$$y = 2x + (-4)$$

n.

$$3(y - 1) = -2(5 - x)$$

$$3y - 3 = -10 + 2x$$

$$3y = -10 + 2x + 3$$

$$3y = 2x - 7$$

$$y = \frac{1}{3}(2x - 7)$$

$$y = \frac{2}{3}x - \frac{7}{3}$$

$$y = \frac{2}{3}x + \left(-\frac{7}{3}\right)$$

o.

$$-(x - y) = 2 - x$$

$$-x + y = 2 - x$$

$$y = 2 - x + x$$

$$y = 2$$

$$y = 0x + 2$$

8. In each solution below, we use the x- and y-intercepts to graph the line unless the intercepts are the same or when there is no x-intercept or y-intercept in which case we will need an additional point.[27] In addition, we use our

---

[27]We will leave it to the reader to find these additional points.

earlier finding that, in the model for a linear relationship, the slope of the corresponding line is m.

a. Formal form:

$$y = 7x + 3$$

x-intercept:

$$y = 7x + 3$$
$$0 = 7x + 3$$
$$7x + 3 = 0$$
$$7x = 0 - 3$$
$$7x = -3$$
$$x = -\frac{3}{7}$$

The x-intercept is $\left(-\frac{3}{7}, 0\right)$.

y-intercept:

$$y = 7x + 3$$
$$y = 7 \times 0 + 3$$
$$y = 0 + 3$$
$$y = 3$$

The y-intercept is $(0, 3)$.

Slope: $m = 7$

Graph:

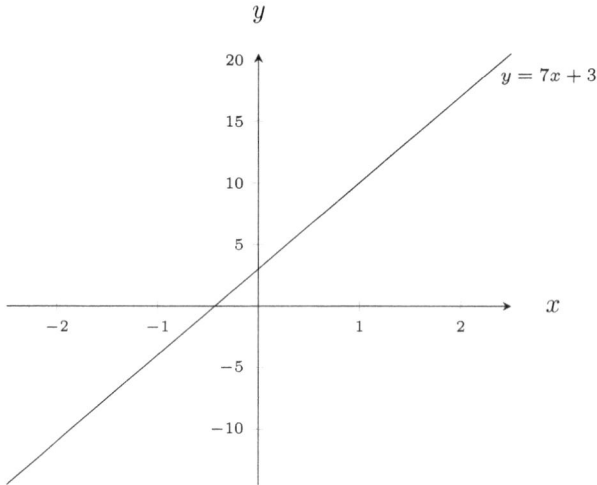

b. Formal form:

$$y = -2x - 1$$
$$y = -2x + (-1)$$

x-intercept:

$$y = -2x - 1$$
$$0 = -2x - 1$$
$$-2x - 1 = 0$$
$$-2x = 0 + 1$$
$$-2x = 1$$
$$x = -\frac{1}{2}$$

The x-intercept is $\left(-\frac{1}{2}, 0\right)$.

y-intercept:

$$y = -2x - 1$$
$$y = -2 \times 0 - 1$$
$$y = 0 - 1$$
$$y = -1$$

The y-intercept is $(0, -1)$.

Slope: $m = -2$

Graph:

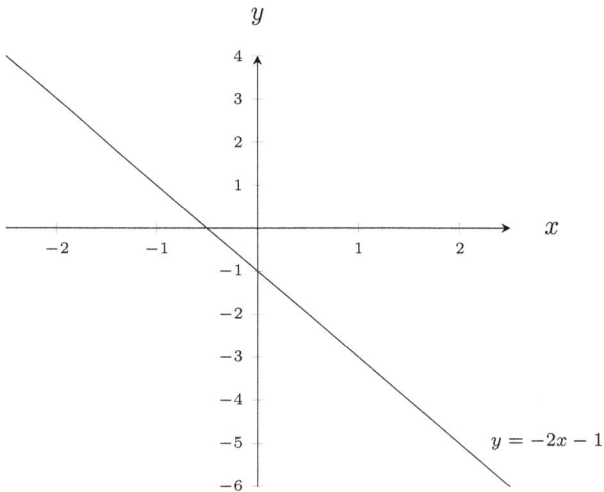

c. Formal form:

$$y = x$$
$$y = 1x + 0$$

x-intercept:

$$y = x$$
$$0 = x$$
$$x = 0$$

The x-intercept is $(0, 0)$.

y-intercept:

$$y = x$$
$$y = 0$$

The y-intercept is $(0, 0)$.

Slope: $m = 1$

Graph:

d. Formal form:

$$x = 3y$$
$$3y = x$$
$$y = \frac{1}{3}x \quad y = \frac{1}{3}x + 0$$

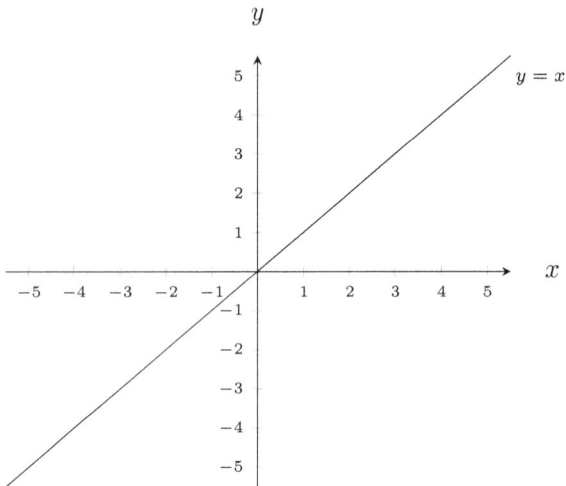

x-intercept:

$$x = 3y$$
$$x = 3 \times 0$$
$$x = 0$$

The x-intercept is $(0, 0)$.

y-intercept:

$$x = 3y$$
$$0 = 3y$$
$$3y = 0$$
$$y = \frac{0}{3}$$
$$y = 0$$

The y-intercept is $(0, 0)$.

Slope: $m = \frac{1}{3}$

Graph:

e. Formal form:

$$y = \frac{3}{4}x$$
$$y = \frac{3}{4}x + 0$$

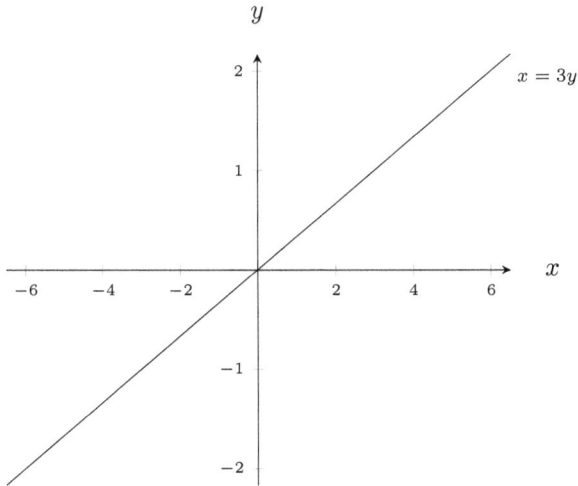

x-intercept:

$$y = \frac{3}{4}x$$

$$0 = \frac{3}{4}x$$

$$\frac{3}{4}x = 0$$

$$x = \frac{4}{3} \times 0$$

$$x = 0$$

The x-intercept is $(0, 0)$.

y-intercept:

$$y = \frac{3}{4}x$$

$$y = \frac{3}{4} \times 0$$

$$y = 0$$

The y-intercept is $(0, 0)$.

Slope: $m = \frac{3}{4}$

Graph:

f. Formal form:

$$\frac{y}{x} = 4$$

$$y = 4x$$

$$y = 4x + 0$$

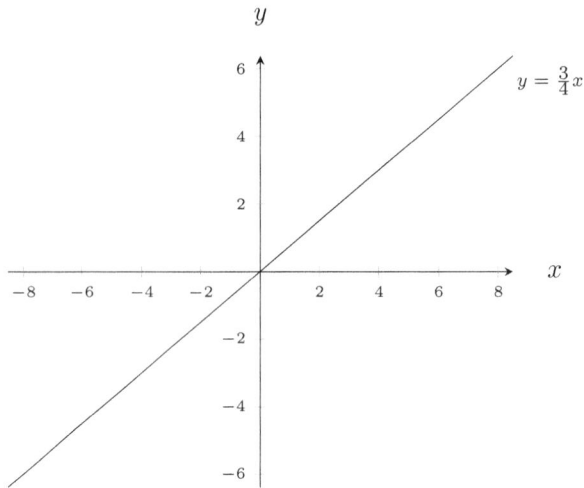

x-intercept:

$$\frac{y}{x} = 4$$
$$y = 4x$$
$$0 = 4x$$
$$4x = 0$$
$$x = \frac{0}{4}$$
$$x = 0$$

The x-intercept is $(0, 0)$.

y-intercept:

$$\frac{y}{x} = 4$$
$$y = 4x$$
$$y = 4 \times 0$$
$$y = 0$$

The y-intercept is $(0, 0)$.

Slope: $m = 4$

Graph:

g. Formal form:

$$\frac{y}{2x} = -1$$
$$y = -2x$$
$$y = -2x + 0$$

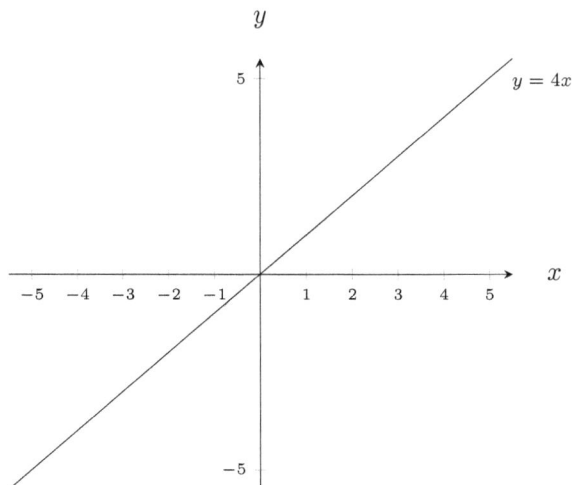

x-intercept:

$$\frac{y}{2x} = -1$$

$$\frac{0}{2x} = -1$$

$$0 = -1 \times 2x$$

$$0 = -2x$$

$$-2x = 0$$

$$x = \frac{0}{-2}$$

$$x = 0$$

The x-intercept is $(0,0)$.

y-intercept:

$$\frac{y}{2x} = -1$$

$$\frac{y}{2 \times 0} = -1$$

$$\frac{y}{0} = -1$$

$$y = 0\,(-1)$$

$$y = 0$$

The y-intercept is $(0,0)$.

Slope: $m = -2$

Graph:

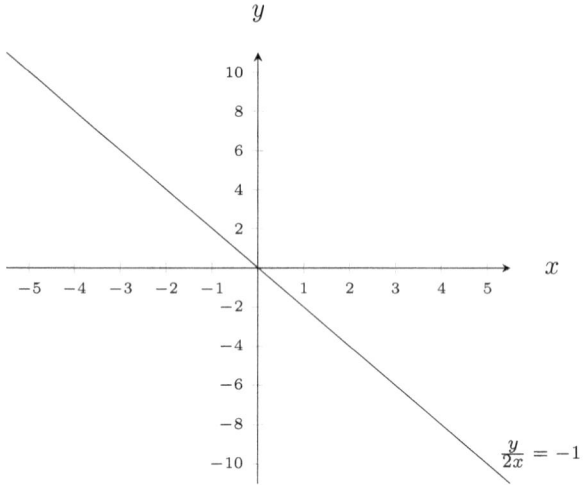

h. Formal form:

$$4y - x = 0$$
$$4y = x + 0$$
$$4y = x$$
$$y = \frac{1}{4}x$$
$$y = \frac{1}{4}x + 0$$

x-intercept:

$$4y - x = 0$$
$$4 \times 0 - x = 0$$
$$0 - x = 0$$
$$-x = 0$$
$$x = 0$$

The x-intercept is $(0,0)$.

y-intercept:

$$4y - x = 0$$
$$4y - 0 = 0$$
$$4y = 0$$
$$y = \frac{0}{4}$$
$$y = 0$$

The y-intercept is $(0,0)$.

Slope: $m = \frac{1}{4}$

Graph:

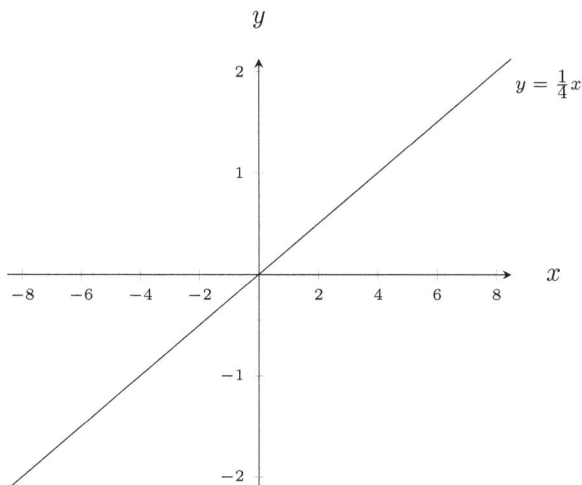

i. Formal form:

$$x = 3y - 7$$
$$3y - 7 = x$$
$$3y = x + 7$$
$$y = \frac{1}{3}(x + 7)$$
$$y = \frac{1}{3}x + \frac{7}{3}$$

x-intercept:

$$x = 3y - 7$$
$$x = 3 \times 0 - 7$$
$$x = 0 - 7$$
$$x = -7$$

The x-intercept is $(-7, 0)$.

y-intercept:

$$x = 3y - 7$$
$$0 = 3y - 7$$
$$3y - 7 = 0$$
$$3y = 0 + 7$$
$$3y = 7$$
$$y = \frac{7}{3}$$

The y-intercept is $\left(0, \frac{7}{3}\right)$.

Slope: $m = \frac{1}{3}$

Graph:

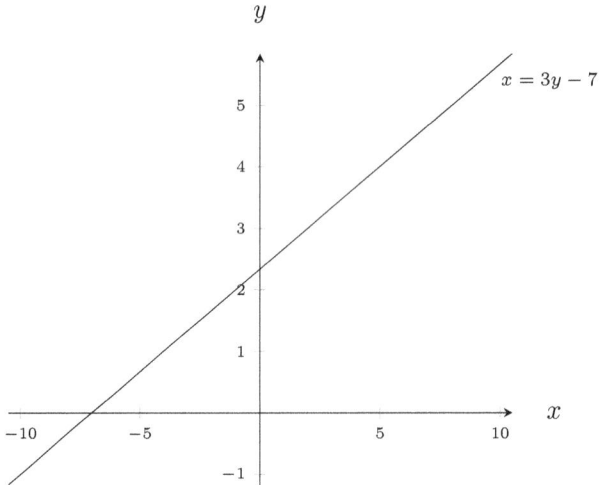

j. Formal form:

$$x + y = 0$$
$$y = -x + 0$$
$$y = -1x + 0$$

x-intercept:

$$x + y = 0$$
$$x + 0 = 0$$
$$x = 0$$

The x-intercept is $(0, 0)$.

y-intercept:

$$x + y = 0$$
$$0 + y = 0$$
$$y = 0$$

The y-intercept is $(0, 0)$.

Slope: $m = -1$

Graph:

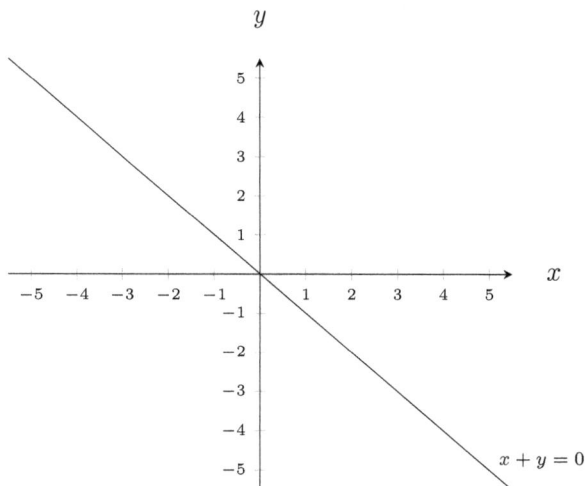

k. Formal form:

$$-\frac{y}{x} = -6$$
$$y = -6\,(-x)$$
$$y = 6x$$
$$y = 6x + 0$$

x-intercept:

$$-\frac{y}{x} = -6$$
$$-\frac{0}{x} = -6$$
$$x = -\frac{0}{-6}$$
$$x = 0$$

The x-intercept is $(0,0)$.

y-intercept:

$$-\frac{y}{x} = -6$$
$$-\frac{y}{0} = -6$$
$$y = -(-6)\,(0)$$
$$y = 0$$

The y-intercept is $(0,0)$.

Slope: $m = 6$

Graph:

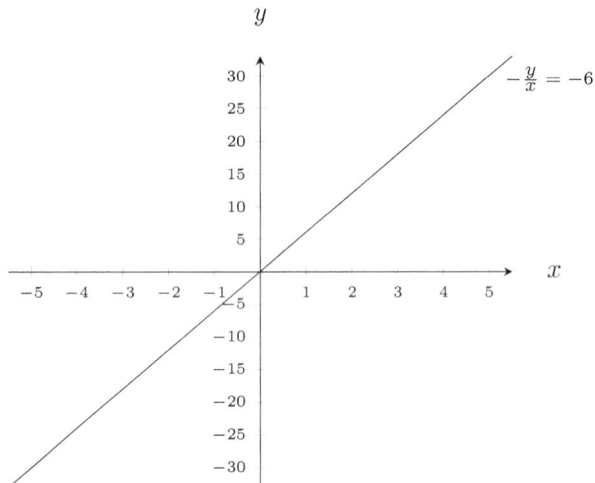

1. Formal form:

$$x = -y$$
$$-y = x$$
$$y = -x$$
$$y = -1x + 0$$

x-intercept:

$$x = -y$$
$$x = -0$$
$$x = 0$$

The x-intercept is $(0, 0)$.

y-intercept:

$$x = -y$$
$$0 = -y$$
$$y = 0$$

The y-intercept is $(0, 0)$.

Slope: $m = -1$

Graph:

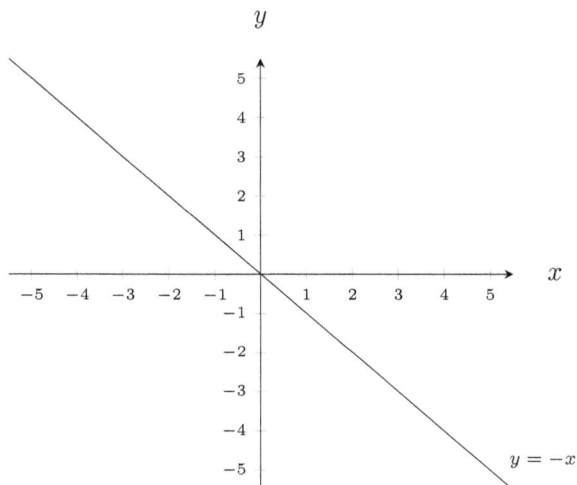

m. Formal form:

$$y = 2(x - 2)$$
$$y = 2x - 4$$

x-intercept:

$$y = 2(x - 2)$$
$$0 = 2(x - 2)$$
$$2(x - 2) = 0$$
$$x - 2 = \frac{0}{2}$$
$$x - 2 = 0$$
$$x = 0 + 2$$
$$x = 2$$

The x-intercept is $(2, 0)$.

y-intercept:

$$y = 2(x - 2)$$
$$y = 2(0 - 2)$$
$$y = 2(-2)$$
$$y = -4$$

The y-intercept is $(0, -4)$.

Slope: $m = 2$

Graph:

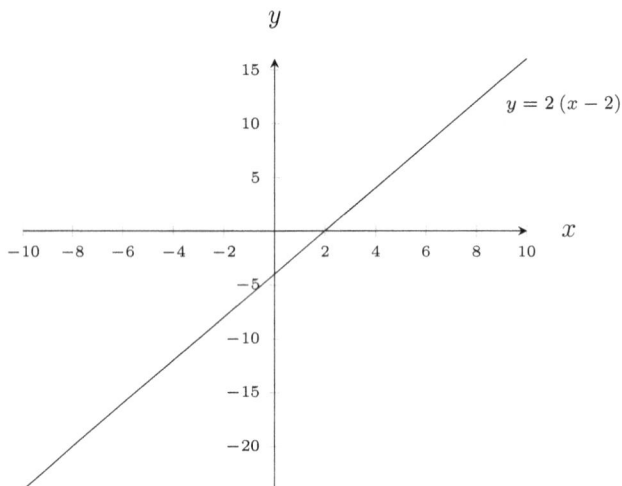

n. Formal form:

$$3\left(y - 1\right) = -2\left(5 - x\right)$$
$$3y - 3 = -10 + 2x$$
$$3y = -10 + 2x + 3$$
$$3y = 2x - 7$$
$$y = \frac{1}{3}\left(2x - 7\right)$$
$$y = \frac{2}{3}x - \frac{7}{3}$$
$$y = \frac{2}{3}x + \left(-\frac{7}{3}\right)$$

x-intercept:

$$3(y - 1) = -2(5 - x)$$
$$3(0 - 1) = -2(5 - x)$$
$$3(-1) = -2(5 - x)$$
$$-3 = -2(5 - x)$$
$$-2(5 - x) = -3$$
$$5 - x = \frac{-3}{-2}$$
$$5 - x = \frac{3}{2}$$
$$-x = \frac{3}{2} - 5$$
$$-x = \frac{3 - 10}{2}$$
$$-x = -\frac{7}{2}$$
$$x = \frac{7}{2}$$

The x-intercept is $\left(\frac{7}{2}, 0\right)$.

y-intercept:

$$3(y - 1) = -2(5 - x)$$
$$3(y - 1) = -2(5 - 0)$$
$$3(y - 1) = -2 \times 5$$
$$3(y - 1) = -10$$
$$y - 1 = \frac{-10}{3}$$
$$y = \frac{-10}{3} + 1$$
$$y = \frac{-10 + 3}{3}$$
$$y = \frac{-7}{3}$$
$$y = -\frac{7}{3}$$

The y-intercept is $\left(0, -\frac{7}{3}\right)$.

Slope: $m = \frac{2}{3}$

Graph:

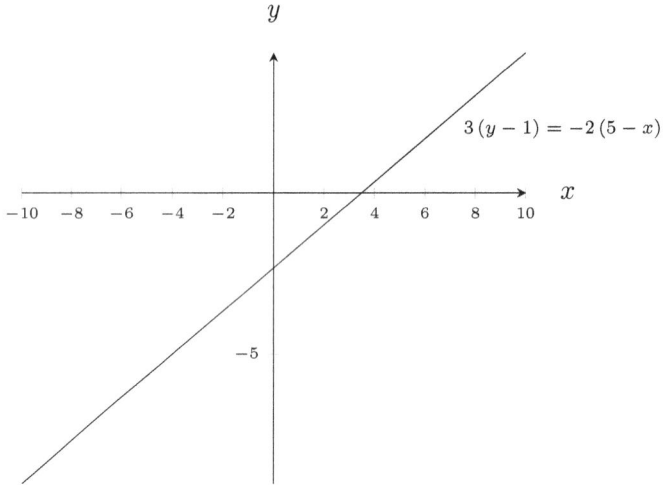

o. Formal form:

$$-(x - y) = 2 - x$$
$$-x + y = 2 - x$$
$$y = 2 - x + x$$
$$y = 2$$
$$y = 0x + 2$$

x-intercept:

$$-(x - y) = 2 - x$$
$$-(x - 0) = 2 - x$$
$$-x = 2 - x$$
$$-x + x = 2$$
$$0x = 2$$
$$x = \frac{2}{0}$$

x is undefined. There are no x-intercepts.

y-intercept:

$$-(x - y) = 2 - x$$
$$-(0 - y) = 2 - 0$$
$$-(-y) = 2$$
$$y = 2$$

The y-intercept is $(0, 2)$.

Slope: $m = 0$

Graph:

$$y$$

$$- (x - y) = 2 - x$$

$$x$$

$$-10 \quad -8 \quad -6 \quad -4 \quad -2 \qquad 2 \quad 4 \quad 6 \quad 8 \quad 10$$

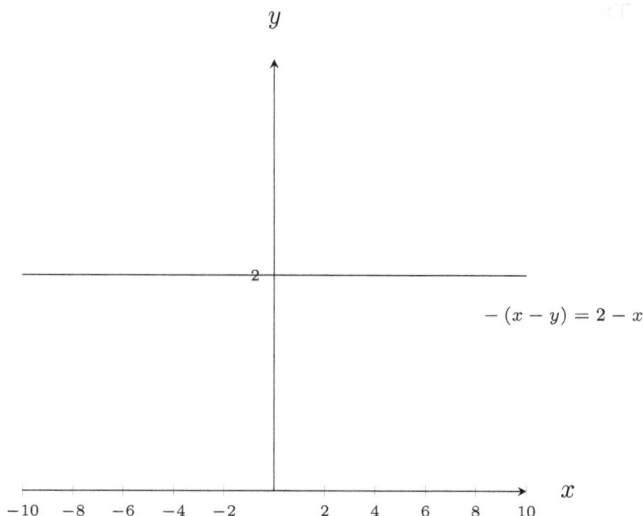

Exercise Set 13.2.1.1

1. Since E_m is directly superpositional to E_k, an increase in E_k by 12 000 J forces an increase in E_m by 12 000 J. For E_m to remain constant, E_p will have to decrease by 12 000 J. Since E_m is directly superpositional to E_p, a decrease in E_p by 12 000 J will force a decrease in E_m by 12 000 J. The combined effect of increasing E_m by 12 000 J and decreasing E_m by 12 000 J is to force a net change in E_m by $12 000 - 12 000$ or 0 J, i.e., E_m does not change.

2. a. Since A is directly superpositional to s^2, a decrease in s^2 by 0.75 m^2 forces a decrease in A by 0.75 m^2. Since A is directly superpositional to $4\pi r^2$, an increase in $4\pi r^2$ by 0.5 m^2 forces an increase in A by 0.5 m^2. The combined effect of decreasing A by 0.75 m^2 and increasing A by 0.5 m^2 is to decrease A by $0.75 - 0.5$ or 0.25 m^2.

   b. Since A is directly superpositional to s^2, an increase in s^2 by 4.2 m^2 forces an increase in A by 4.2 m^2. To force A to increase by 3.5 m^2, $4\pi r^2$ will have to decrease by 0.7 m^2. Since A is directly superpositional to $4\pi r^2$, a decrease in $4\pi r^2$ by 0.7 m^2 forces a decrease in A by 0.7 m^2. The combined effect of increasing A by 4.2 m^2 and decreasing A by 0.7 m^2 is to increase A by $4.2 - 0.7$ or 3.5 m^2.

Exercise Set 13.2.1.2

1. a. The expressions on the left and right sides of the formula, $A = \frac{1}{2}bh$, analyze into one term each. Therefore, the formula is a proportion formula.

   b. The relationship between A and b is direct proportion. This is because the factors appear on opposite sides of the formula and are both being multiplied by.

The relationship between A and h is also direct proportion. This is because the factors appear on opposite sides of the formula and are both being multiplied by.

The relationship between b and h is inverse proportion. This is because the factors appear on the same side of the formula and are both being multiplied by.

c. Since the relationship between A and b is direct proportion, an increase in b by a factor of 2 forces an increase in A by a factor of 2. Since h does not change, it will have no effect on the change in A.

d. Since the relationship between A and h is direct proportion, a decrease in h by a factor of $\frac{1}{2}$ forces a decrease in A by a factor of $\frac{1}{2}$. Since b does not change, it will have no effect on the change in A.

e. Since the relationship between A and b is direct proportion, an increase in b by a factor of 2 forces an increase in A by a factor of 2. Since the relationship between A and h is direct proportion, a decrease in h by a factor of $\frac{1}{2}$ forces a decrease in A by a factor of $\frac{1}{2}$. The combined effect of increaseing A by a factor of 2 and decreasing A by a factor of $\frac{1}{2}$ is to force a change in A by a factor of $2 \times \frac{1}{2}$ or 1 in A, i.e., A does not change.

f. Since the relationship between A and b is direct proportion, an increase in b by a factor of 2 forces an increase in A by a factor of 2. Since the relationship between A and h is direct proportion, we can increase h by a factor of 5 which forces an increase in A by a factor of 5. The combined effect of increasing A by a factor of 2 and increasing A by a factor of 5 is to increase A by a factor of 2×5 or 10, i.e., A becomes 10 times larger.

2. a. The expressions on the left and right sides of the formula, $E = \frac{1}{2}mv^2$, analyze into one term each. Therefore, the formula is a proportion formula.

b. The relationship between E and m is direct proportion. This is because the factors appear on opposite sides of the formula and are both being multiplied by.

The relationship between E and v is more complex than either superposition or proportion. This is because v does not appear as a single term or a single factor.

The relationship between m and v is also more complex than either superposition or proportion. This is because they do not represent terms and v is not a factor.

c. Since the relationship between E and m is direct proportion, an increase in m by a factor of 2 forces an increase in E by a factor of 2 as well. Since v does not change, it will have no effect on the change in E.

d. Since the relationship between E and v is one of power, an increase in v by a factor of 3 forces an increase in E by a factor of 3^2 or 9. Since m does not change, it will have no effect on the change in E.

e. Since the relationship between E and m is direct proportion, an increase in m by a factor of 2 forces an increase in E by a factor of 2. Since the relationship between E and v is of type power, a decrease in v by a factor of $\frac{1}{2}$ forces a decrease in E by a factor or $\left(\frac{1}{2}\right)^2$ or $\frac{1}{4}$. The combined effect of increasing E by a factor of 2 and decreasing E by a factor of $\frac{1}{4}$, is to force a decrease in E by a factor of $2 \times \frac{1}{4}$ or $\frac{1}{2}$, i.e., E halves.

f. Since the relationship between E and m is direct proportion, an increase in E by a factor of 2 forces an increase in m by a factor of 2 as well. Since the relationship between m and v^2 is inverse proportion, a decrease in v^2 by a factor of $\frac{1}{8}$ will force an increase in m by a factor of 8. The combined effect of increasing m by a factor of 2 and increasing m by a factor of 8 is to increase m by a factor of 2×8 or 16.

To decrease v^2 by a factor of $\frac{1}{8}$, we can decrease v by a factor of $\sqrt{\frac{1}{8}}$ which is equal to a factor of about 0.354.

3. a. The expressions on the left and right sides of the formula, $V = lwh$, analyze into one term each. Therefore, the formula is a proportion formula.
   b. The relationship between V and l is direct proportion. This is because the factors appear on opposite sides of the formula and are both being multiplied by.

   The relationship between V and w is direct proportion. This is because the factors appear on opposite sides of the formula and are both being multiplied by.

   The relationship between V and h is direct proportion. This is because the factors appear on opposite sides of the formula and are both being multiplied by.

   The relationship between l and w is inverse proportion. This is because the factors appear on the same side of the formula and are both being multiplied by.

   The relationship between l and h is inverse proportion. This is because the factors appear on the same side of the formula and are both being multiplied by.

   The relationship between w and h is inverse proportion. This is because the factors appear on the same side of the formula and are both being multiplied by.
   c. Since V is directly proportional to l, an increase in l by a factor of 2 forces an increase in V by a factor of 2, i.e., V doubles. Since w and h do not change, they will have no effect on the change in V.
   d. Since V is directly proportional to w, an increase in w by a factor of 2 forces an increase in V by a factor of 2. Since V is directly proportional to h, an increase in h by a factor of 3 forces an increase in V by a factor of 3. The combined effect of increasing V by a factor of 2 and increasing V by a factor of 3 is to force an increase in V by a factor of 2×3 or 6. Since l does not change, it will have no effect on the change in V.
   e. Since V is directly proportional to w, a decrease in w by a factor of $\frac{1}{2}$ forces a decrease in V by a factor of $\frac{1}{2}$. Since V is directly proportional to h, an increase in h by a factor of 3 forces an increase in V by a factor of 3. The combined effect of decreasing V by a factor of $\frac{1}{2}$ and increasing V by a factor of 3 is to force an increase in V by a factor of $\frac{1}{2} \times 3$ or $\frac{3}{2}$.[28] To keep V the same, we will have to decrease l by a factor of $\frac{2}{3}$. Since V is directly proportional to l, a decrease in l by a factor of $\frac{2}{3}$ forces a decrease in V by a factor of $\frac{2}{3}$. The combined effect of decreasing V by a factor of $\frac{1}{2}$, increasing V by a factor of 3 and decreasing V by a factor of $\frac{2}{3}$ is to force a change in V by a factor of $\frac{1}{2} \times 3 \times \frac{2}{3}$ or 1, i.e., V does not change.

---

[28]This means that V increases by a factor of $1\frac{1}{2}$.

4. a. The expressions on the left and right sides of the formula, $pV = nRT$, analyze into one term each. Therefore, the formula is a proportion formula.

   b. In describing the relatinships between variables we have ignored R as R is a constant.

   The relationship between p and V is inverse proportion. This is because the factors appear on the same side of the formula and are both being multiplied by.

   The relationship between p and n is direct proportion. This is because the factors appear on opposite sides of the formula and are both being multiplied by.

   The relationship between p and T is direct proportion. This is because the factors appear on opposite sides of the formula and are both being multiplied by.

   The relationship between V and n is direct proportion. This is because the factors appear on opposite sides of the formula and are both being multiplied by.

   The relationship between V and T is direct proportion. This is because the factors appear on opposite sides of the formula and are both being multiplied by.

   The relationship between n and T is inverse proportion. This is because the factors appear on the same side of the formula and are both being multiplied by.

   c. Since p is directly proportional to T, an increase in T by a factor of 2 increases p by a factor of 2, i.e., p doubles. Since n and V do not change, they will have no effect on the change in p.

   d. Since p is directly proportional to T, an increase in T by a factor of 3 forces and increase in p by a factor of 3. Since p is inversely proportional to V, an increase in V by a factor of 3 forces a decrease in p by a factor of $\frac{1}{3}$. The combined effect of increasing p by a factor of 3 and decreasing p by a factor of $\frac{1}{3}$ is to force a change in p by a factor of $3 \times \frac{1}{3}$ or 1, i.e., p does not change.

   e. Since p is directly proportional to T, a decrease in T by a factor of $\frac{1}{2}$ forces a decrease in p by a factor of $\frac{1}{2}$. Since p is directly proportional to n, an increase in n by a factor of 4 forces an increase in p by a factor of 4. The combined effect of decreasing p by a factor of $\frac{1}{2}$ and increasing p by a factor of 4 is to force an increase in p by a factor of $\frac{1}{2} \times 4$ or 2, i.e., p doubles.

Exercise Set 13.2.2.1

1. In each case, to show linear dependency, we rearrange the formula and map it onto the linear equation $y = a \pm bx$.

   a. d_0 is linearly dependent on d:

   $$d_0 = d - vt$$
   $$d_0 = -vt + d$$
   $$d_0 = -vt + 1d$$

This has the form of the linear equation $y = a \pm bx$ with $a = -vt$ and $b = 1$.

d_0 is linearly dependent on v:

$$d_0 = d - vt$$
$$d_0 = d - tv$$

This has the form of the linear equation $y = a \pm bx$ with $a = d$ and $b = t$.

d_0 is linearly dependent on t: The equation $d_0 = d - vt$ has the form of the linear equation $y = a \pm bx$ with $a = d$ and $b = v$.

b. m is linearly dependent on m_0:

$$m = \frac{1}{2^n} m_0$$
$$m = 0 + \frac{1}{2^n} m_0$$

This has the form of the linear equation $y = a \pm bx$ with $a = 0$ and $b = \frac{1}{2^n}$.

c. e is linearly dependent on g:

$$e = \frac{1}{2} v^2 + gh$$
$$e = \frac{1}{2} v^2 + hg$$

This has the form of the linear equation $y = a \pm bx$ with $a = \frac{1}{2} v^2$ and $b = h$.

e is linearly dependent on h: The equation $e = \frac{1}{2} v^2 + gh$ has the form of the linear equation $y = a \pm bx$ with $a = \frac{1}{2} v^2$ and $b = g$.

d. m_{CO_2} is linearly dependent on m_C:

$$m_{CO_2} = m_C + 2m_O$$
$$m_{CO_2} = 2m_O + m_C$$
$$m_{CO_2} = 2m_O + 1m_C$$

This has the form of the linear equation $y = a \pm bx$ with $a = 2m_O$ and $b = 1$.

m_{CO_2} is linearly dependent on m_O: The equation $m_{CO_2} = m_C + 2m_O$ has the form of the linear equation $y = a \pm bx$ with $a = m_C$ and $b = 2$.

Exercise Set 13.3

While we follow Step 1 in the algorithm for rearranging formulas given in the textbook, we do not show it in the solutions given below. This makes for a leaner presentation and is a style that we recommend.

In the solutions below, we also make use of both division and its equivalent interpretation as *fraction of* (e.g., $\frac{x}{2}$ and $\frac{1}{2}x$), sometimes using one, sometimes using the other to point out that both interpretations are valid and that both are routinely used. Only in cases where the use of the former leads to the generation of a complex fraction do we use the latter to avoid having to simplify such fractions.

1. a. i.

$$g = \frac{GM}{R^2}$$

$$G = \frac{gR^2}{M}$$

ii.

$$g = \frac{GM}{R^2}$$

$$M = \frac{gR^2}{G}$$

iii.

$$g = \frac{GM}{R^2}$$

$$R^2 = \frac{GM}{g}$$

$$R = \pm\sqrt{\frac{GM}{g}}$$

$$R = \sqrt{\frac{GM}{g}}$$

b. i.

$$a = \frac{v}{t}$$

$$v = at$$

ii.

$$a = \frac{v}{t}$$

$$t = \frac{v}{a}$$

c. i.

$$m_1 v_1 = m_2 v_2$$

$$m_1 = \frac{v_2}{v_1} m_2$$

ii.

$$m_1 v_1 = m_2 v_2$$

$$v_1 = \frac{m_2}{m_1} v_2$$

iii.

$$m_1 v_1 = m_2 v_2$$

$$m_2 = \frac{v_1}{v_2} m_1$$

iv.

$$m_1 v_1 = m_2 v_2$$

$$v_2 = \frac{m_1}{m_2} v_1$$

d. i.

$$T = \frac{1}{f}$$

$$f = \frac{1}{T}$$

e. i.

$$d = \frac{1}{2} a t^2$$

$$a = \frac{2d}{t^2}$$

ii.

$$d = \frac{1}{2} a t^2$$

$$t^2 = \frac{2d}{a}$$

$$t = \pm\sqrt{\frac{2d}{a}}$$

f. i.

$$F = ma$$

$$m = \frac{F}{a}$$

ii.

$$F = ma$$

$$a = \frac{F}{m}$$

g.   i.

$$a = \frac{v^2}{r}$$
$$v^2 = ar$$
$$v = \pm\sqrt{ar}$$
$$v = \sqrt{ar}$$

ii.

$$a = \frac{v^2}{r}$$
$$r = \frac{v^2}{a}$$

h.   i.

$$F = \frac{Gm_1 m_2}{d^2}$$
$$G = \frac{Fd^2}{m_1 m_2}$$

ii.

$$F = \frac{Gm_1 m_2}{d^2}$$
$$m_1 = \frac{Fd^2}{Gm_2}$$

iii.

$$F = \frac{Gm_1 m_2}{d^2}$$
$$m_2 = \frac{Fd^2}{Gm_1}$$

iv.

$$F = \frac{Gm_1 m_2}{d^2}$$
$$d^2 = \frac{Gm_1 m_2}{F}$$
$$d = \pm\sqrt{\frac{Gm_1 m_2}{F}}$$
$$d = \sqrt{\frac{Gm_1 m_2}{F}}$$

i.   i.

$$E_k = \frac{1}{2}mv^2$$
$$m = \frac{2E_k}{v^2}$$

ii.

$$E_k = \frac{1}{2}mv^2$$
$$v^2 = \frac{2E_k}{m}$$
$$v = \pm\sqrt{\frac{2E_k}{m}}$$
$$v = \sqrt{\frac{2E_k}{m}}$$

j.   i.

$$d = \frac{v^2}{2g}$$
$$v^2 = 2gd$$
$$v = \pm\sqrt{2gd}$$
$$v = \sqrt{2gd}$$

ii.

$$d = \frac{v^2}{2g}$$
$$g = \frac{v^2}{2d}$$

k.   i.

$$P = \frac{W}{t}$$
$$W = Pt$$

ii.

$$P = \frac{W}{t}$$
$$t = \frac{W}{P}$$

l.   i.

$$V = lwh$$
$$l = \frac{V}{wh}$$

ii.

$$V = lwh$$
$$w = \frac{V}{lh}$$

iii.
$$V = lwh$$
$$h = \frac{V}{lw}$$

m.   i.
$$\nu = f\lambda$$
$$f = \frac{\nu}{\lambda}$$

ii.
$$\nu = f\lambda$$
$$\lambda = \frac{\nu}{f}$$

n.   i.
$$V = IR$$
$$I = \frac{V}{R}$$

ii.
$$V = IR$$
$$R = \frac{V}{I}$$

o.   i.
$$\frac{V_0}{V_i} = \frac{N_0}{N_i}$$
$$V_0 = \frac{N_0}{N_i}V_i$$

ii.
$$\frac{V_0}{V_i} = \frac{N_0}{N_i}$$
$$V_i = \frac{N_i}{N_0}V_0$$

iii.
$$\frac{V_0}{V_i} = \frac{N_0}{N_i}$$
$$N_0 = \frac{V_0}{V_i}N_i$$

iv.
$$\frac{V_0}{V_i} = \frac{N_0}{N_i}$$
$$N_i = \frac{V_i}{V_0}N_0$$

p.   i.
$$M = \frac{-p}{s}$$
$$p = -sM$$

ii.
$$M = \frac{-p}{s}$$
$$s = \frac{-p}{M}$$

q.   i.
$$\lambda = \frac{h}{mv}$$
$$h = \lambda mv$$

ii.
$$\lambda = \frac{h}{mv}$$
$$m = \frac{h}{\lambda v}$$

iii.
$$\lambda = \frac{h}{mv}$$
$$v = \frac{h}{\lambda m}$$

r.   i.
$$E = mc^2$$
$$m = \frac{E}{c^2}$$

ii.
$$E = mc^2$$
$$c^2 = \frac{E}{m}$$
$$c = \pm\sqrt{\frac{E}{m}}$$
$$c = \sqrt{\frac{E}{m}}$$

s.   i.
$$m = \frac{1}{2^n}m_0$$
$$2^n = \frac{m_0}{m}$$
$$n = \log_2\frac{m_0}{m}$$

ii.
$$m = \frac{1}{2^n}m_0$$
$$m_0 = 2^n m$$

2. a. i.

$$a = \frac{v - v_0}{t - t_0}$$
$$v - v_0 = a(t - t_0)$$
$$v = v_0 + a(t - t_0)$$

ii.

$$a = \frac{v - v_0}{t - t_0}$$
$$v - v_0 = a(t - t_0)$$
$$-v_0 = -v + a(t - t_0)$$
$$v_0 = v - a(t - t_0)$$

iii.

$$a = \frac{v - v_0}{t - t_0}$$
$$t - t_0 = \frac{v - v_0}{a}$$
$$t = t_0 + \frac{v - v_0}{a}$$

iv.

$$a = \frac{v - v_0}{t - t_0}$$
$$t - t_0 = \frac{v - v_0}{a}$$
$$-t_0 = -t + \frac{v - v_0}{a}$$
$$t_0 = t - \frac{v - v_0}{a}$$

b. i.

$$F = \frac{mv - m_0 v_0}{t - t_0}$$
$$mv - m_0 v_0 = F(t - t_0)$$
$$mv = m_0 v_0 + F(t - t_0)$$
$$m = \frac{1}{v}\left[m_0 v_0 + F(t - t_0)\right]$$

ii.

$$F = \frac{mv - m_0 v_0}{t - t_0}$$
$$mv - m_0 v_0 = F(t - t_0)$$
$$mv = m_0 v_0 + F(t - t_0)$$
$$v = \frac{1}{m}\left[m_0 v_0 + F(t - t_0)\right]$$

iii.

$$F = \frac{mv - m_0 v_0}{t - t_0}$$
$$mv - m_0 v_0 = F(t - t_0)$$
$$-m_0 v_0 = -mv + F(t - t_0)$$
$$m_0 v_0 = mv - F(t - t_0)$$
$$m_0 = \frac{1}{v_0}\left[mv - F(t - t_0)\right]$$

iv.

$$F = \frac{mv - m_0 v_0}{t - t_0}$$
$$mv - m_0 v_0 = F(t - t_0)$$
$$-m_0 v_0 = -mv + F(t - t_0)$$
$$m_0 v_0 = mv - F(t - t_0)$$
$$v_0 = \frac{1}{m_0}\left[mv - F(t - t_0)\right]$$

v.

$$F = \frac{mv - m_0 v_0}{t - t_0}$$
$$t - t_0 = \frac{mv - m_0 v_0}{F}$$
$$t = t_0 + \frac{1}{F}(mv - m_0 v_0)$$

vi.

$$F = \frac{mv - m_0 v_0}{t - t_0}$$
$$t - t_0 = \frac{mv - m_0 v_0}{F}$$
$$-t_0 = -t + \frac{mv - m_0 v_0}{F}$$
$$t_0 = t - \frac{mv - m_0 v_0}{F}$$

c. i.

$$d = v_0 t + \frac{1}{2}at^2$$
$$v_0 t + \frac{1}{2}at^2 = d$$
$$v_0 t = d - \frac{1}{2}at^2$$
$$v_0 = \frac{1}{t}\left(d - \frac{1}{2}at^2\right)$$

ii.

$$d = v_0 t + \frac{1}{2} a t^2$$

$$v_0 t + \frac{1}{2} a t^2 = d$$

$$\frac{1}{2} a t^2 = d - v_0 t$$

$$a = \frac{2(d - v_0 t)}{t^2}$$

d.   i.

$$d = d_0 + v_0 t + \frac{1}{2} a t^2$$

$$d_0 + v_0 t + \frac{1}{2} a t^2 = d$$

$$d_0 = d - v_0 t - \frac{1}{2} a t^2$$

ii.

$$d = d_0 + v_0 t + \frac{1}{2} a t^2$$

$$d_0 + v_0 t + \frac{1}{2} a t^2 = d$$

$$v_0 t = d - d_0 - \frac{1}{2} a t^2$$

$$v_0 = \frac{1}{t} \left(d - d_0 - \frac{1}{2} a t^2 \right)$$

iii.

$$d = d_0 + v_0 t + \frac{1}{2} a t^2$$

$$d_0 + v_0 t + \frac{1}{2} a t^2 = d$$

$$\frac{1}{2} a t^2 = d - d_0 - v_0 t$$

$$a = \frac{2(d - d_0 - v_0 t)}{t^2}$$

e.   i.

$$p = 2(l + w)$$

$$l + w = \frac{1}{2} p$$

$$l = \frac{1}{2} p - w$$

ii.

$$p = 2(l + w)$$

$$l + w = \frac{1}{2} p$$

$$w = \frac{1}{2} p - l$$

f.   i.

$$A = \frac{1}{2} (b_1 + b_2) h$$

$$b_1 + b_2 = \frac{2A}{h}$$

$$b_1 = \frac{2A}{h} - b_2$$

ii.

$$A = \frac{1}{2} (b_1 + b_2) h$$

$$b_1 + b_2 = \frac{2A}{h}$$

$$b_2 = \frac{2A}{h} - b_1$$

iii.

$$A = \frac{1}{2} (b_1 + b_2) h$$

$$h = \frac{2A}{b_1 + b_2}$$

g.   i.

$$l_2 - l_1 = \alpha l_1 (T_2 - T_1)$$

$$l_2 = l_1 + \alpha l_1 (T_2 - T_1)$$

ii.

$$l_2 - l_1 = \alpha l_1 (T_2 - T_1)$$

$$l_2 = l_1 + \alpha l_1 (T_2 - T_1)$$

$$l_1 + \alpha l_1 (T_2 - T_1) = l_2$$

$$\left[1 + \alpha (T_2 - T_1) \right] l_1 = l_2$$

$$l_1 = \frac{l_2}{1 + \alpha (T_2 - T_1)}$$

iii.

$$l_2 - l_1 = \alpha l_1 (T_2 - T_1)$$

$$\alpha l_1 (T_2 - T_1) = l_2 - l_1$$

$$\alpha = \frac{l_2 - l_1}{l_1 (T_2 - T_1)}$$

iv.

$$l_2 - l_1 = \alpha l_1 (T_2 - T_1)$$

$$T_2 - T_1 = \frac{l_2 - l_1}{\alpha l_1}$$

$$T_2 = T_1 + \frac{l_2 - l_1}{\alpha l_1}$$

v.

$$l_2 - l_1 = \alpha l_1 (T_2 - T_1)$$

$$T_2 - T_1 = \frac{l_2 - l_1}{\alpha l_1}$$

$$-T_1 = -T_2 + \frac{l_2 - l_1}{\alpha l_1}$$

$$T_1 = T_2 - \frac{l_2 - l_1}{\alpha l_1}$$

h. i.

$$Q = mC(T_2 - T_1)$$

$$m = \frac{Q}{C(T_2 - T_1)}$$

ii.

$$Q = mC(T_2 - T_1)$$

$$C = \frac{Q}{m(T_2 - T_1)}$$

iii.

$$Q = mC(T_2 - T_1)$$

$$T_2 - T_1 = \frac{Q}{mC}$$

$$T_2 = T_1 + \frac{Q}{mC}$$

iv.

$$Q = mC(T_2 - T_1)$$

$$T_2 - T_1 = \frac{Q}{mC}$$

$$-T_1 = -T_2 + \frac{Q}{mC}$$

$$T_1 = T_2 - \frac{Q}{mC}$$

i. i.

$$p = \frac{sf}{s - f}$$

$$p(s - f) = sf$$

$$ps - pf = sf$$

$$ps - sf = pf$$

$$(p - f)s = pf$$

$$s = \frac{pf}{p - f}$$

ii.

$$p = \frac{sf}{s - f}$$

$$p(s - f) = sf$$

$$ps - pf = sf$$

$$ps = sf + pf$$

$$sf + pf = ps$$

$$(s + p)f = ps$$

$$f = \frac{ps}{s + p}$$

j. i.

$$x_2 - x_1 = \frac{S}{a}\lambda$$

$$x_2 = x_1 + \frac{S}{a}\lambda$$

ii.

$$x_2 - x_1 = \frac{S}{a}\lambda$$

$$-x_1 = -x_2 + \frac{S}{a}\lambda$$

$$x_1 = x_2 - \frac{S}{a}\lambda$$

iii.

$$x_2 - x_1 = \frac{S}{a}\lambda$$

$$S = \frac{a(x_2 - x_1)}{\lambda}$$

iv.

$$x_2 - x_1 = \frac{S}{a}\lambda$$

$$a = \frac{S\lambda}{x_2 - x_1}$$

v.

$$x_2 - x_1 = \frac{S}{a}\lambda$$

$$\lambda = \frac{a(x_2 - x_1)}{S}$$

k. i.

$$E_2 - E_1 = hf$$

$$E_2 = E_1 + hf$$

ii.

$$E_2 - E_1 = hf$$
$$- E_1 = -E_2 + hf$$
$$E_1 = E_2 - hf$$

iv.

$$E_2 - E_1 = hf$$
$$f = \frac{E_2 - E_1}{h}$$

iii.

$$E_2 - E_1 = hf$$
$$h = \frac{E_2 - E_1}{f}$$

Exercise Set I.1

1. A tabular prime factorization of the denominators 5, 6 and 3 leads to the least common denominator of $2 \times 3 \times 5$ or 30.

2. A tabular prime factorization of the denominators 4, 5 and 3 leads to the least common denominator of $2 \times 2 \times 3 \times 5$ or 60.

3. A tabular prime factorization of the denominators 8, 4 and 2 leads to the least common denominator of $2 \times 2 \times 2$ or 8.

4. A tabular prime factorization of the denominators 12, 8 and 6 leads to the least common denominator of $2 \times 2 \times 2 \times 3$ or 24.

5. A tabular prime factorization of the denominators 12, 3, 5 and 6 leads to the least common denominator of $2 \times 2 \times 3 \times 5$ or 60.

6. A tabular prime factorization of the denominators 15, 12 and 8 leads to the least common denominator of $2 \times 2 \times 2 \times 3 \times 5$ or 120.

7. A tabular prime factorization of the denominators 17, 34 and 2 leads to the least common denominator of 2×17 or 34.

8. A tabular prime factorization of the denominators 14, 21 and 7 leads to the least common denominator of $2 \times 3 \times 7$ or 42.

9. A tabular prime factorization of the denominators 16, 4 and 3 leads to the least common denominator of $2 \times 2 \times 2 \times 2 \times 3$ or 48.

10. A tabular prime factorization of the denominators 25, 15 and 3 leads to the least common denominator of $3 \times 5 \times 5$ or 75.

11. A tabular prime factorization of the denominators 10, 2, 4 and 8 leads to the least common denominator of $2 \times 2 \times 2 \times 5$ or 40.

12. A tabular prime factorization of the denominators 18, 15 and 12 leads to the least common denominator of $2 \times 2 \times 3 \times 3 \times 5$ or 180.

13. A tabular prime factorization of the denominators 7, 5 and 3 leads to the least common denominator of $3 \times 5 \times 7$ or 105.

14. A tabular prime factorization of the denominators 22, 33 and 11 leads to the least common denominator of $2 \times 3 \times 11$ or 66.

15. A tabular prime factorization of the denominators 19, 2, 38 and 3 leads to the least common denominator of $2 \times 3 \times 19$ or 114.

Exercise Set I.2

1. $\frac{12}{30}, \frac{25}{30}, \frac{10}{30}$

2. $\frac{15}{60}, \frac{24}{60}, \frac{40}{60}$

3. $\frac{3}{8}, \frac{2}{8}, \frac{4}{8}$

4. $\frac{10}{24}, \frac{3}{24}, \frac{4}{24}$

5. $\frac{5}{60}, \frac{40}{60}, \frac{12}{60}, \frac{50}{60}$

6. $\frac{8}{120}, \frac{50}{120}, \frac{75}{120}$

7. $\frac{2}{34}, \frac{5}{34}, \frac{51}{34}$

8. $\frac{27}{42}, \frac{4}{42}, \frac{18}{42}$

9. $\frac{9}{48}, \frac{12}{48}, \frac{32}{48}$

10. $\frac{12}{75}, \frac{5}{75}, \frac{50}{75}$

11. $\frac{28}{40}, \frac{20}{40}, \frac{30}{40}, \frac{5}{40}$

12. $\frac{10}{180}, \frac{36}{180}, \frac{15}{180}$

13. $\frac{165}{105}, \frac{21}{105}, \frac{35}{105}$

14. $\frac{6}{66}, \frac{4}{66}, \frac{90}{66}$

15. $\frac{18}{114}, \frac{57}{114}, \frac{12}{114}, \frac{76}{114}$

Exercise Set M

1. a.
$$x^6 x^7 = x^{6+7}$$
$$= x^{13}$$

b.
$$x^{19} x^{-3} = x^{19-3}$$
$$= x^{16}$$

c.
$$x^0 x^{-1} = x^{0-1}$$
$$= x^{-1}$$
$$= \frac{1}{x^1}$$
$$= \frac{1}{x}$$

d.
$$y^{-5} y^5 = y^{-5+5}$$
$$= y^0$$
$$= 1$$

e.
$$x^{-4} x^{-12} = x^{-4-12}$$
$$= x^{-16}$$
$$= \frac{1}{x^{16}}$$

f.
$$x^{-5} x^6 x^0 = x^{-5+6+0}$$
$$= x^1$$
$$= x$$

g.
$$x^2 x^0 x^{-1} = x^{2+0-1}$$
$$= x^1$$
$$= x$$

h.
$$z^{-3} z^{-3} z^{-3} = z^{-3-3-3}$$
$$= z^{-9}$$
$$= \frac{1}{z^9}$$

2. a.
$$\frac{x^8}{x^2} = x^{8-2}$$
$$= x^6$$

b.
$$\frac{x^{-17}}{x^{-12}} = x^{-17+12}$$
$$= x^{-5}$$
$$= \frac{1}{x^5}$$

c.

$$\frac{x^2}{x^0} = x^{2-0}$$
$$= x^2$$

d.

$$\frac{y^{-3}}{y^{-5}} = y^{-3+5}$$
$$= y^2$$

3. a.

$$\frac{x^5 x^4}{x^7} = x^{5+4-7}$$
$$= x^2$$

b.

$$\frac{x^{-10}}{x^8 x^5} = x^{-10-8-5}$$
$$= x^{-23}$$
$$= \frac{1}{x^{23}}$$

c.

$$\frac{x^3 x^{-1}}{x^6} = x^{3-1-6}$$
$$= x^{-4}$$
$$= \frac{1}{x^4}$$

d.

$$\frac{y^{-3} y^{-2}}{y^{-5} y^{-1}} = y^{-3-2+5+1}$$
$$= y^1$$
$$= y$$

e.

$$\frac{z^0}{z^3 z^7} = z^{0-3-7}$$
$$= z^{-10}$$
$$= \frac{1}{z^{10}}$$

f.

$$\frac{x^{-4} x^{-7}}{x^{-6}} = x^{-4-7+6}$$
$$= x^{-5}$$
$$= \frac{1}{x^5}$$

g.

$$\frac{x^2}{x^0 x^{-3}} = x^{2-0+3}$$
$$= x^5$$

h.

$$\frac{z^{-1} z^0}{z^{-2} z^5 z^2} = z^{-1+0+2-5-2}$$
$$= z^{-6}$$
$$= \frac{1}{z^6}$$

4. a.

$$\left(x^3\right)^6 = x^{3\times6}$$
$$= x^{18}$$

b.

$$\left(x^6\right)^3 = x^{6\times3}$$
$$= x^{18}$$

c.

$$\left(x^{-7}\right)^3 = x^{-7\times3}$$
$$= x^{-21}$$
$$= \frac{1}{x^{21}}$$

d.
$$\left(x^{-1}\right)^5 = x^{-1\times 5}$$
$$= x^{-5}$$
$$= \frac{1}{x^5}$$

g.
$$\left(x^2\right)^8 = x^{2\times 8}$$
$$= x^{16}$$

e.
$$\left(x^0\right)^{-2} = x^{0(-2)}$$
$$= x^0$$
$$= 1$$

h.
$$\left(x^{-7}\right)^2 = x^{-7\times 2}$$
$$= x^{-14}$$
$$= \frac{1}{x^{14}}$$

f.
$$\left(x^5\right)^5 = x^{5\times 5}$$
$$= x^{25}$$

5. a.
$$x^{3^4} = x^{81}$$

b.
$$x^{2^5} = x^{32}$$

c.
$$x^{6^1} = x^6$$

d.
$$x^{(-3)^2} = x^9$$

Exercise Set N.2

1. a. $\frac{57}{100}$ d. $\frac{2}{25}$ g. $\frac{217}{800}$ j. $\frac{1}{120}$

   b. $\frac{1}{5}$ e. $\frac{1}{100}$ h. $\frac{11}{400}$ k. $\frac{3}{800}$

   c. $\frac{8}{25}$ f. $\frac{99}{100}$ i. $\frac{177}{500}$ l. $\frac{9}{1000}$

2. a. 57%   d. 8%   g. $27\frac{1}{8}\%$ j. $\frac{5}{6}\%$

   b. 20%   e. 1%   h. $2\frac{3}{4}\%$ k. $\frac{3}{8}\%$

   c. 32%   f. 99%   i. $35\frac{2}{5}\%$ l. $\frac{9}{10}\%$

Exercise Set N.3

1. a. 0.57   e. 0.01   i. 0.142   m. 3.509

   b. 0.2   f. 0.99   j. 0.096   n. 0.0002

   c. 0.32   g. 0.457   k. 0.024   o. 0.005

   d. 0.08   h. 0.239   l. 1.205   p. 0.000 04

2. a. 57%   f. 99%   k. 0.9%   p. 2.4%

   b. 20%   g. 27.125%   l. 45.7%   q. 120.5%

   c. 32%   h. 2.75%   m. 23.9%   r. 350.9%

   d. 8%   i. 3.54%   n. 14.2%   s. 0.02%

   e. 1%   j. 0.375%   o. 9.6%   t. 0.004%

Exercise Set O.2

1. A$(2,5)$, B$(0,4)$, C$(4,-6)$, D$(-8,6)$, E$(-5,-2)$, F$(-3,0)$, G$(0,-5)$, H$(9,0)$
2. A$(1.5,4.3)$, B$(2.6,-2.1)$, C$(-4.2,-3.8)$, D$(-3.5,0)$
3. $(2,-1.8)$, $(-3,2.9)$, $(-4.9,0)$, $(4.4,3.7)$, $(0,-2.7)$

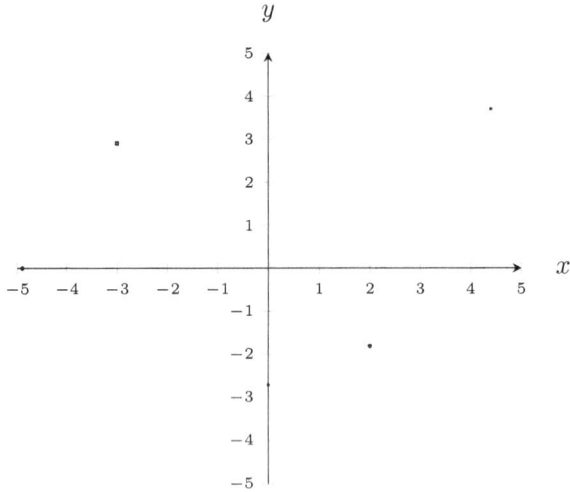

4.

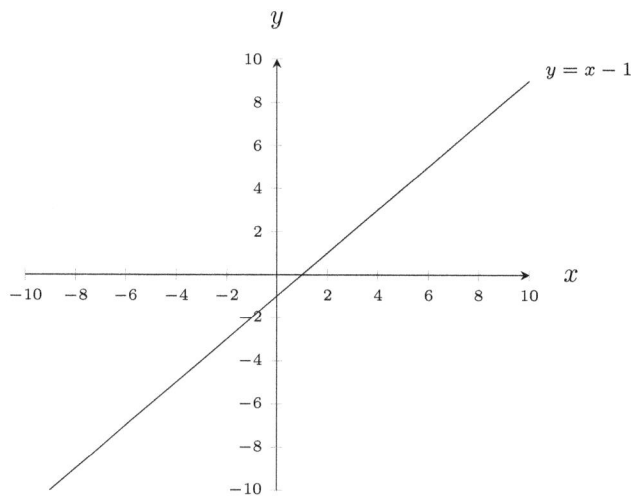

$y = x - 1$

a.

b.

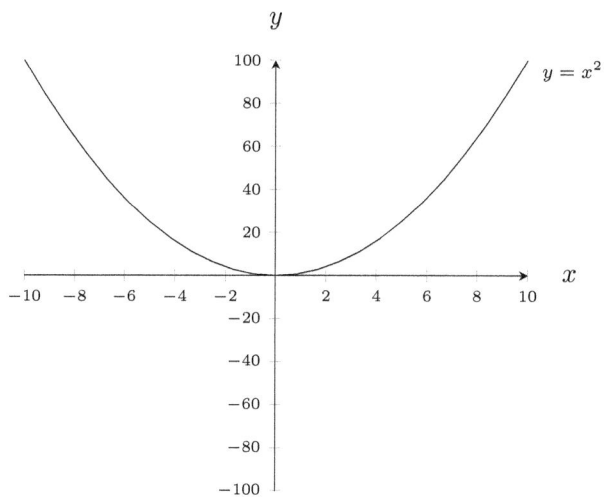

$y = x^2$

c.

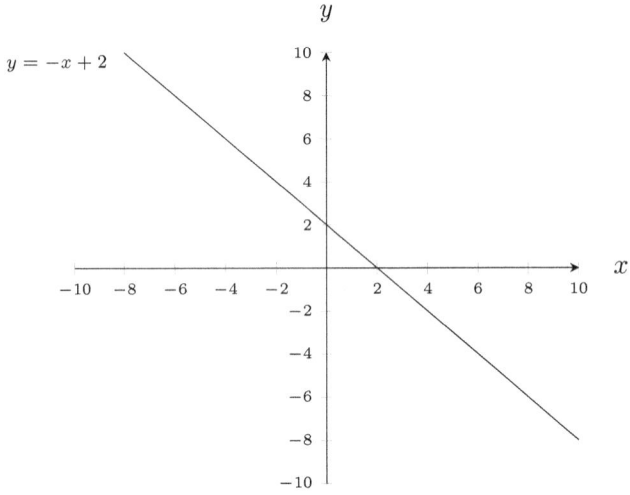

The graph shows a line labeled $y = -x + 2$ on an xy-coordinate plane with axes marked from -10 to 10.

d.

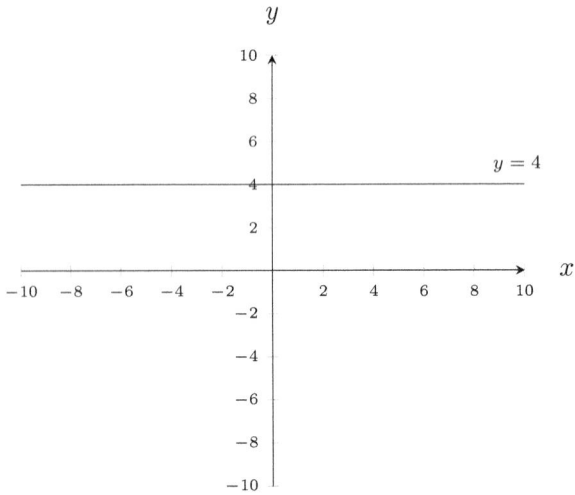

The graph shows a horizontal line labeled $y = 4$ on an xy-coordinate plane with axes marked from -10 to 10.

e.

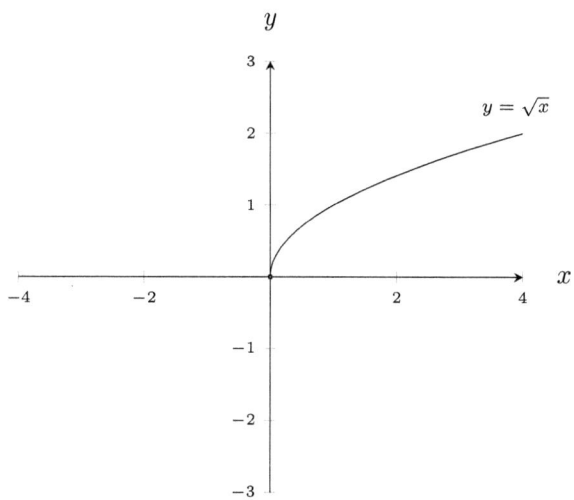

$y = \sqrt{x}$

f.

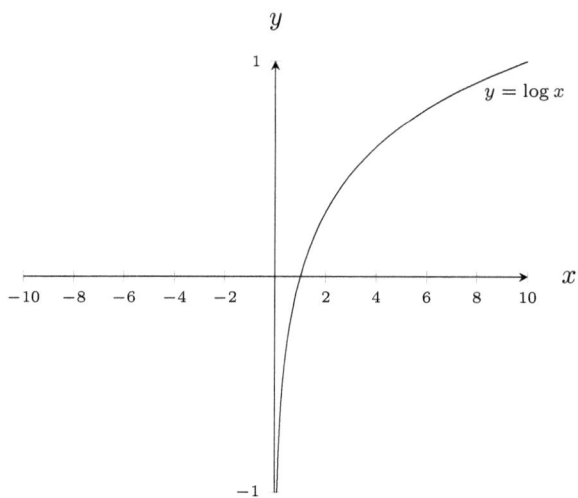

$y = \log x$

g.

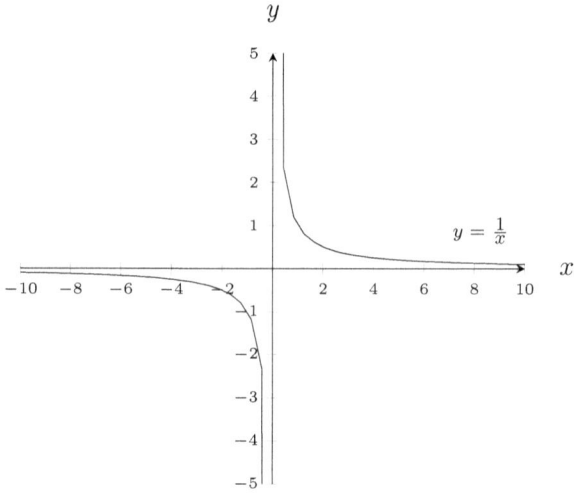

h.

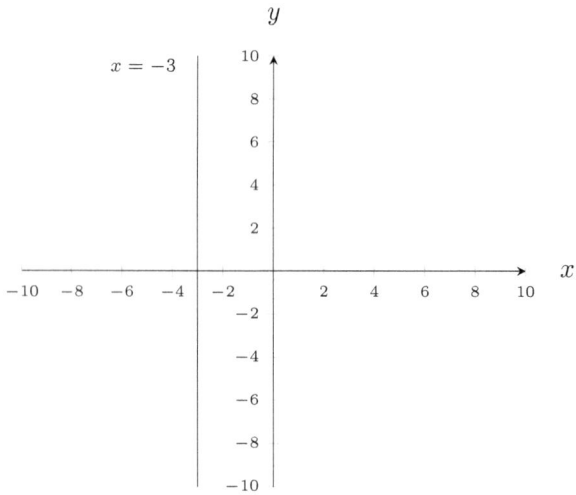

5.

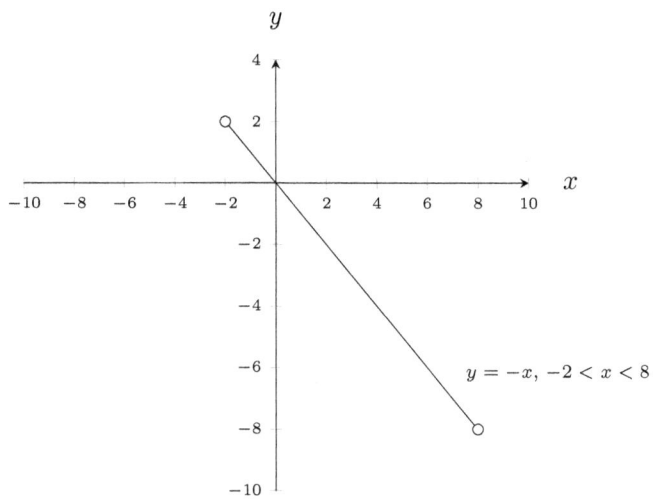

$$y = -x, \; -2 < x < 8$$

a.

b.

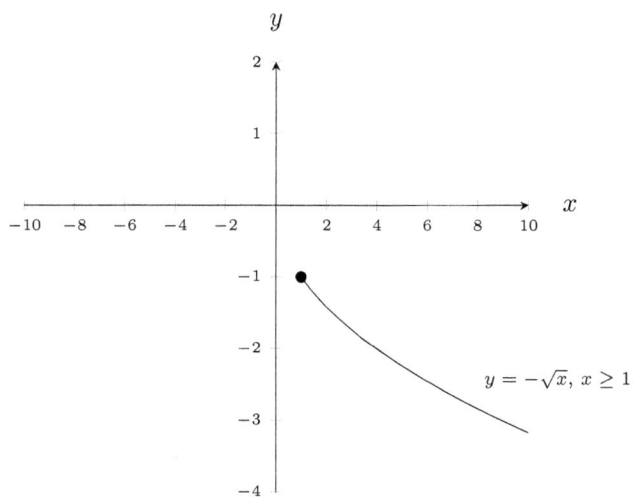

$$y = -\sqrt{x}, \; x \geq 1$$

c.

$y = x^2 + 2, \ x < -1$

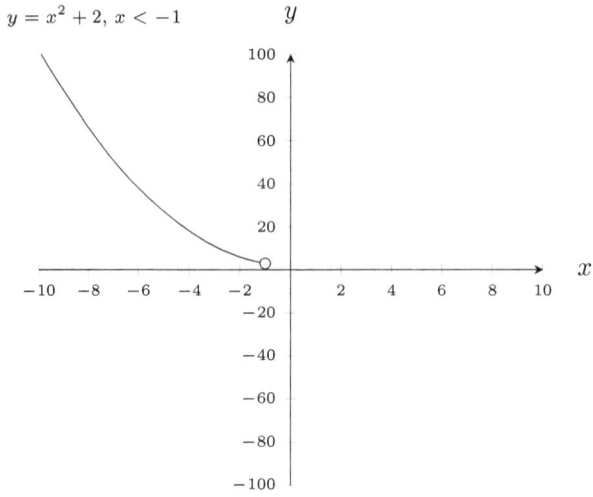

d.

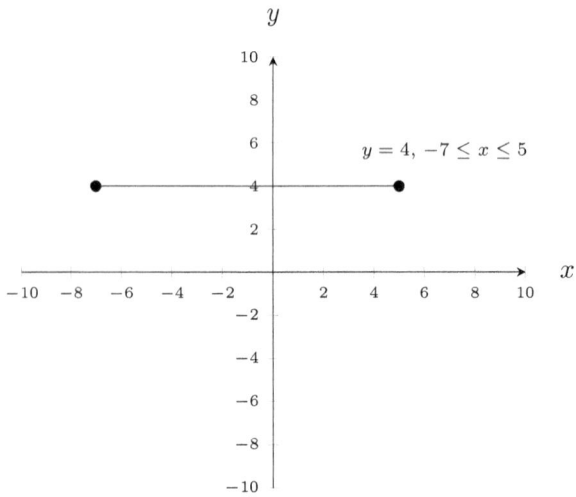

$y = 4, \ -7 \le x \le 5$

e.

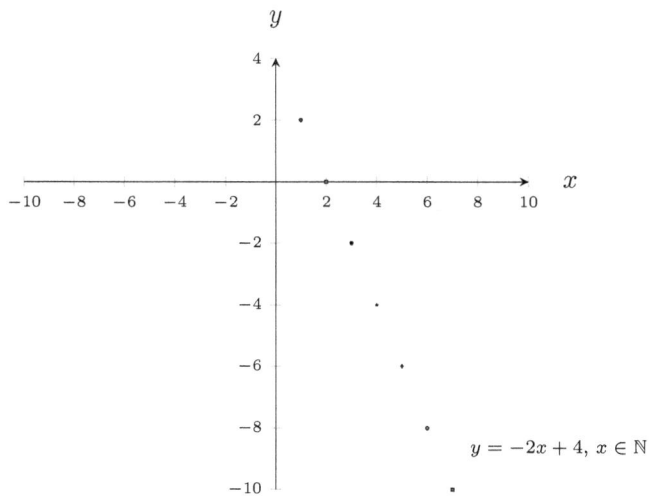

$$y = -2x + 4,\ x \in \mathbb{N}$$

f.

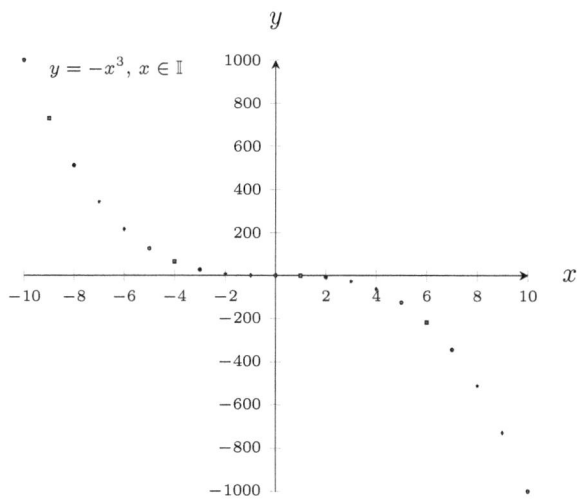

$$y = -x^3,\ x \in \mathbb{I}$$

Exercise Set O.3

1.

x-intercept: $(1,0)$
y−intercept: $(0,-1)$

b. x-intercept: $(2,0)$
y-intercept: $(0,2)$

c. x-intercept: $(2,0)$
y-intercept: $(0,-1)$

d. No x-intercept

y-intercept: $(0,6)$

e. No x-intercept
No y-intercept

f. x-intercept: $(6,0)$
y-intercept: $(0,-3)$

g. x-intercepts: $(-7.2,0)$ and $(7.2,)$
y-intercept: $(0,-6)$

2. In each case we set y equal to 0 to find the x-intercept, and we set x equal to 0 to find the y-intercept.

a. x-intercepts:

$$y = 4x + 3$$
$$0 = 4x + 3$$
$$4x + 3 = 0$$
$$4x = 0 - 3$$
$$4x = -3$$
$$x = -\frac{3}{4}$$

x-intercept: $\left(-\frac{3}{4}, 0\right)$

y-intercepts:

$$y = -x + 3$$
$$y = -0 + 3$$
$$y = 3$$

y-intercept: $(0,3)$

y-intercepts:

$$y = 4x + 3$$
$$y = 4 \times 0 + 3$$
$$y = 0 + 3$$
$$y = 3$$

y-intercept: $(0,3)$

c. x-intercepts:

$$y - x = 4$$
$$0 - x = 4$$
$$-x = 4$$
$$x = -4$$

x-intercept: $(-4,0)$

b. x-intercepts:

$$y = -x + 3$$
$$0 = -x + 3$$
$$-x + 3 = 0$$
$$-x = 0 - 3$$
$$-x = -3$$
$$x = 3$$

x-intercept: $(3,0)$

y-intercepts:

$$y - x = 4$$
$$y - 0 = 4$$
$$y = 4$$

y-intercept: $(0,4)$

d. x-intercepts:

$$y = -2$$
$$0 = -2$$

No x-intercept

More formally:

$$y = -2$$
$$y + 0x = -2$$
$$0 + 0x = -2$$
$$0x = -2$$
$$x = \frac{-2}{0}$$

x is undefined

No x-intercept

y-intercepts:

$$y = -2$$

y-intercept: $(0, -2)$

e. x-intercepts:

$$y = x^2 - 1$$
$$0 = x^2 - 1$$
$$x^2 - 1 = 0$$
$$x^2 = 0 + 1$$
$$x^2 = 1$$
$$x = \pm\sqrt{1}$$
$$x = \pm 1$$

x-intercepts: $(-1, 0)$ and $(1, 0)$

y-intercepts:

$$y = x^2 - 1$$
$$y = 0^2 - 1$$
$$y = 0 - 1$$
$$y = -1$$

y-intercept: $(0, -1)$

f. x-intercepts:

$$y = x^3$$
$$0 = x^3$$
$$x^3 = 0$$
$$x = \sqrt[3]{0}$$
$$x = 0$$

x-intercept: $(0, 0)$

y-intercepts:

$$y = x^3$$
$$y = 0^3$$
$$y = 0$$

y-intercept: $(0, 0)$

g. x-intercepts:

$$y = -\frac{3}{x}$$
$$0 = -\frac{3}{x}$$
$$x = -\frac{3}{0}$$
$$x = \frac{-3}{0}$$

x is undefined

No x-intercept

y-intercepts:

$$y = -\frac{3}{x}$$
$$y = -\frac{3}{0}$$
$$y = \frac{-3}{0}$$

y is undefined

No y-intercept

h. x-intercepts:

$$x^2 + y^2 = 4$$
$$x^2 + 0^2 = 4$$
$$x^2 + 0 = 4$$
$$x^2 = 4$$
$$x = \pm\sqrt{4}$$
$$x = \pm 2$$

x-intercepts: $(2,0)$ and $(-2,0)$

y-intercepts:

$$x^2 + y^2 = 4$$
$$0^2 + y^2 = 4$$
$$0 + y^2 = 4$$
$$y^2 = 4$$
$$y = \pm\sqrt{4}$$
$$y = \pm 2$$

y-intercepts: $(0,2)$ and $(0,-2)$

Exercise Set O.4

1. a.

$$m = \frac{\Delta y}{\Delta x}$$
$$m = \frac{3}{1}$$
$$m = 3$$

b.

$$m = \frac{\Delta y}{\Delta x}$$
$$m = \frac{6}{2}$$
$$m = 3$$

c.

$$m = \frac{\Delta y}{\Delta x}$$
$$m = \frac{-3}{1}$$
$$m = -3$$

d.

$$m = \frac{\Delta y}{\Delta x}$$
$$m = \frac{-6}{2}$$
$$m = -3$$

e.

$$m = \frac{\Delta y}{\Delta x}$$
$$m = \frac{3}{-1}$$
$$m = -3$$

f.

$$m = \frac{\Delta y}{\Delta x}$$
$$m = \frac{6}{-2}$$
$$m = -3$$

g.

$$m = \frac{\Delta y}{\Delta x}$$
$$m = \frac{-3}{-1}$$
$$m = 3$$

h.

$$m = \frac{\Delta y}{\Delta x}$$
$$m = \frac{-6}{-2}$$
$$m = 3$$

2. a.

$$m = \frac{\Delta y}{\Delta x}$$

$$m = \frac{4.7}{2.3}$$

$$m = 2.043 \cdots$$

b.

$$m = \frac{\Delta y}{\Delta x}$$

$$m = \frac{5.8}{0.1}$$

$$m = 58$$

c.

$$m = \frac{\Delta y}{\Delta x}$$

$$m = \frac{-7.1}{20.3}$$

$$m = -0.349 \cdots$$

d.

$$m = \frac{\Delta y}{\Delta x}$$

$$m = \frac{0}{3}$$

$$m = 0$$

3. a.

$$m = \frac{y_2 - y_1}{x_2 - x_1}$$

$$m = \frac{7 - 2}{4 - 3}$$

$$m = \frac{5}{1}$$

$$m = 5$$

b.

$$m = \frac{y_2 - y_1}{x_2 - x_1}$$

$$m = \frac{-1 - 4}{2 - (-3)}$$

$$m = \frac{-5}{2 + 3}$$

$$m = \frac{-5}{5}$$

$$m = -1$$

c.

$$m = \frac{y_2 - y_1}{x_2 - x_1}$$

$$m = \frac{0 - 0}{0 - 5.2}$$

$$m = \frac{0}{-5.2}$$

$$m = 0$$

d.

$$m = \frac{y_2 - y_1}{x_2 - x_1}$$

$$m = \frac{2 - 5}{-3 - (-3)}$$

$$m = \frac{-3}{-3 + 3}$$

$$m = \frac{-3}{0}$$

m is undefined

e.

$$m = \frac{y_2 - y_1}{x_2 - x_1}$$

$$m = \frac{9.9 - 2.1}{4.7 - 8.4}$$

$$m = \frac{7.8}{-3.7}$$

$$m = -2.108 \cdots$$

f.

$$m = \frac{y_2 - y_1}{x_2 - x_1}$$

$$m = \frac{2 - 2}{-5 - 3}$$

$$m = \frac{0}{-8}$$

$$m = 0$$

g.

$$m = \frac{y_2 - y_1}{x_2 - x_1}$$

$$m = \frac{-\frac{1}{2} - \frac{2}{5}}{\frac{3}{4} - \frac{1}{3}}$$

$$m = \frac{\frac{-5 - 4}{10}}{\frac{9 - 4}{12}}$$

$$m = \frac{\frac{-9}{10}}{\frac{5}{12}}$$

$$m = -\frac{9 \times 12}{10 \times 5}$$

$$m = -\frac{54}{25}$$

h.

$$m = \frac{y_2 - y_1}{x_2 - x_1}$$

$$m = \frac{-1.7 - (-2.0)}{-5 - 3.6}$$

$$m = \frac{-1.7 + 2}{-8.6}$$

$$m = \frac{0.3}{-8.6}$$

$$m = -0.034\,883 \cdots$$

4. $45°$
5. $0°$
6. $-45°$

7. The angle that the line makes with the horizontal is less than $-45°$.

8. The angle that the line makes with the horizontal is less than $45°$.

www.ingramcontent.com/pod-product-compliance
Lightning Source LLC
Chambersburg PA
CBHW060005210326
41520CB00009B/824